Introduction to Mechanical Vibrations

Introduction to Mechanical Vibrations

Ronald J. Anderson
Queen's University
Kingston
Canada

This edition first published 2020

© 2020 John Wiley & Sons Ltd

The right of Ronald J. Anderson to be identified as the author of this work has been asserted in accordance with law.

Registered Offices

John Wiley & Sons, Inc., 111 River Street, Hoboken, NJ 07030, USA

John Wiley & Sons Ltd, The Atrium, Southern Gate, Chichester, West Sussex, PO19 8SQ, UK

Editorial Office

The Atrium, Southern Gate, Chichester, West Sussex, PO19 8SQ, UK

For details of our global editorial offices, customer services, and more information about Wiley products visit us at www.wiley.com.

Wiley also publishes its books in a variety of electronic formats and by print-on-demand. Some content that appears in standard print versions of this book may not be available in other formats.

Library of Congress Cataloging-in-Publication data applied for

HB ISBN: 9781119053651

Cover Design: Wiley

Cover Image: © mammuth/Getty Images

Set in 9.5/12.5pt STIXTwoText by SPi Global, Chennai, India

Printed and bound by CPI Group (UK) Ltd, Croydon, CR0 4YY

10 9 8 7 6 5 4 3 2 1

To June

Contents

Preface

When I first studied vibrations, as an undergraduate student, its importance was clear to our class because it was a required course for mechanical engineers. A few years later, when I started teaching vibrations and new topics were entering the field of mechanical engineering, a course on vibrations was no longer seen as being important enough to be a required so it became an elective. Now, although "mechanical engineering" is still used as an umbrella term, the students who graduate are mechanical engineers with a specialization. Students in the specialized streams do not have time to cover all of the topics that used to be expected of mechanical engineers so some graduate without thermodynamics, others without vibrations, and so on. Specialization like this is inevitable given the expanding scope of knowledge in engineering and the limited time available to undergraduate students but it means even fewer students are learning about vibrations and other important topics. While preparing this introduction to vibrations, I kept in mind the need for undergraduate students to have a better understanding of two topics that are ubiquitous in today's engineering workplace – finite element analysis (FEA) and fast Fourier transforms (FFT). FEA and FFT software tools are readily available to both students and practicing engineers and they need to be used with understanding and a degree of caution.

I was never able to find a textbook that covered just enough, and the right, material for a semester length introductory course in vibrations. I used many textbooks over the years but there was never a fit with what I thought should be in an introduction to vibrations. I was looking for something student-friendly in that it should be readable, almost conversational, but still be mathematically rigorous. What I found on the market were mainly "reference books" as opposed to "teaching books". Many of the textbooks I tried are very good at covering, in depth, a broad range of topics in vibrations, but students have difficulty using them as a first text in the subject, mainly because of the overwhelming amount of material presented.

This book grew from my attempt to accomplish two things in a single course in vibrations. The primary goal is, of course, to present the basics of vibrations in a manner that promotes understanding and interest while building a foundation of knowledge in the field. To do this, I have had to give only brief coverage of many important topics with the hope that some students will go on to expand their knowledge in these areas if their interest is piqued. As mentioned earlier, a secondary goal is to give students a good understanding of finite element analysis and Fourier transforms. While these two subjects fit nicely into vibrations, this book presents them in a way that emphasizes understanding of the underlying principles so that students are aware of both the power and the limitations of the methods.

Chapter 1 addresses the way in which a student has to think about previous undergraduate dynamics knowledge in order to make the transition to analysis of vibrating systems. It introduces the idea of small motions about a stable equilibrium state and addresses the details of linearization. Lagrange's Equations are introduced here and students take to them very quickly as an alternative to Newton's Laws.

Chapter 2 considers the details of analyzing single degree of freedom systems. While much of this material is obvious to those skilled in vibrations, it is vital material for developing the students' abilities. It covers topics such as preloads in springs and why gravitational forces don't need to be included because they are canceled out by the constant preloads. It looks at the constitutive relationship for a spring and shows how to draw free body diagrams consistently and accurately.

Chapter 3 is about free vibrations of single degree of freedom systems. It covers systems with and without damping and tries to make sense of what it means to solve a second-order, linear differential equation without being too prescriptive about it.

Chapter 4 looks at time response when applying a harmonic forcing function to an undamped single degree of freedom system, thereby introducing the phenomena of beating and resonance. This is a short chapter although the subject of time response, if presented in detail, could make for a very long one. Time response is an area that I see as being of secondary importance in an introduction to vibrations.

Chapter 5 considers steady state forced vibrations, covering harmonic forcing functions, harmonic base motion, systems with a rotating unbalance, and accelerometer design. This is really the essence of vibrational analysis and is covered in detail.

Chapter 6 is devoted to the very important subject of damping. Linear viscous damping is discussed and the concept of modeling other energy removal devices as "equivalent linear viscous dampers" is introduced. Coulomb damping is covered. The concept of logarithmic decrement is introduced.

Chapter 7 recognizes that systems often have more than one degree of freedom. Deriving the equations of motion for systems with many degrees of freedom is discussed. The concept of multiple natural frequencies, each associated with a different mode shape, is covered in detail. Description of mode shapes is given a lot of time because of its importance in the field. Forced vibrations, vibration absorbers, and the method of normal modes are covered.

Chapter 8 moves on into the study of continuous systems and uses vibrations of a taut string and a cantilever beam as examples of two continuous systems where solutions can be found. The concept of infinitely many degrees of freedom is introduced.

Chapter 9 recognizes that solutions cannot always be found for continuous systems so the finite element method is introduced as an alternate way to get solutions. Shape functions, element mass and stiffness matrices, and assembly of global mass and stiffness matrices are covered, as well as application of boundary conditions and applied forces. Derivations here are handled using Lagrange's Equation because the students are familiar with that approach by the time we get to finite elements. This is certainly not the approach taken by experts in finite elements but it is a useful and appropriate way to get the students to understand the assumptions made in using FEA.

Chapter 10 is devoted to a relatively new device called the "inerter". This device provides a force that is proportional to the relative acceleration across it. A concept for how to construct such a device and analyses showing its effects on vibrations are presented. This is presented to give the students a look at developing technology with which they can have some fun while realizing that pat answers coming from what they have learned about vibrations to this point may not apply to this device.

Chapter 11 presents a detailed description of how to analyze experimental data in studying vibrations. Topics covered include: Discrete Fourier transforms; sampling and aliasing; leakage and windowing. The approach I have taken here is very old-fashioned in that it treats the Discrete Fourier Transform as something that is derived from a least squares curve fit. This harks back to methods used more than fifty years ago but it enhances students' understanding of what lies behind the transformation to the frequency domain. This is a long chapter that presents methods that the students can program themselves so they don't have to be tied down to packaged FFT programs. Once they have their software working, they can experiment with data sets that clearly demonstrate aliasing, leakage, and so on.

Chapter 12 discusses a variety of topics in vibrations such as how to handle the mass of a spring, flow-induced vibrations, self-excited vibrations in rail vehicles, rigid body modes, and things you can determine from the static deflection of a system. I find that I am usually able to get somewhere into this chapter before the semester is over. The material in this chapter is interesting but certainly doesn't need to be covered to have a complete introduction to vibrations.

It is my hope that this book strikes the right balance for professors teaching introductory vibrations and for their students. I wish them all well.

September, 2019   *Ronald J. Anderson*
Kingston, Canada

About the Companion Website

The companion website for this book is at

www.wiley.com/go/anderson/introduction-to-vibrations

The website includes:

- Animated GIFs
- PDFs and
- Modes software files

Scan this QR code to visit the companion website.

1

The Transition from Dynamics to Vibrations

Introductory undergraduate courses on dynamics typically consider large scale motions of systems of particles and/or rigid bodies and instantaneous solutions to their nonlinear, governing equations. You may recall working on dynamics problems where a system of bodies starts from rest at a prescribed position and your task was to determine, for example, the angular acceleration of a body or the forces acting on some part of the system. Solutions like this, while having some utility, provide only part of the understanding of the system that is required for a successful design. In most cases, the derived governing equations are complete enough but the "snapshot" solutions don't help much with the design process.

There are, in fact, many things that can be done with the equations governing the dynamic motion of the system. Briefly, they can be used to

1. Find where the bodies in the system would be if the system were at rest. These are the *Equilibrium States.*
2. Determine whether the equilibrium states are stable or unstable.
3. Determine how the system behaves for small motions away from a stable equilibrium state.
4. Determine the response of the system in the time domain through the use of numerical simulations. This is the most complex type of analysis and, perhaps surprisingly, gives the least information to the designer until the design has reached the fine tuning phase. The simulations are the analog of "cut and try" experiments where an unsuccessful result gives little information on what to change in order to improve the design.

While going through the material presented in this book, you will be concentrating on very small motions of systems about stable equilibrium states. In doing so, you will see connections to topics you may have covered in courses on statics, on dynamics, and on control systems. You will become very familiar with the linearized, differential, equations of motion for dynamic systems moving around stable equilibrium states and methods for deriving and solving them. This is the essence of *Vibrations.*

To get started and as a review of sorts we begin with the dynamic analysis to a relatively simple system – a bead sliding on a rotating semicircular wire.

Introduction to Mechanical Vibrations, First Edition. Ronald J. Anderson.
© 2020 John Wiley & Sons Ltd. Published 2020 by John Wiley & Sons Ltd.
Companion website: www.wiley.com/go/anderson/introduction-to-vibrations

1.1 Bead on a Wire: The Nonlinear Equations of Motion

First courses on the subject of Dynamics, whether for particles or rigid bodies, are primarily concerned with teaching the basics of kinematics, free body diagrams, and applications of Newton's Laws of Motion. Applying these three concepts sequentially will lead to a set of simultaneous force and moment balance equations that take account of kinematic constraints.

There are different ways of approaching these problems. One can use a formal vector-based approach and we will start with that here because it gives a complete set of governing equations including solutions for all constraint forces that are required to enforce kinematic constraints on the motion. A shorthand version of this approach which may be called an "informal vector approach" is often used in practice and that will be the second method addressed here. It typically works with two-dimensional views and leads to the governing equations of motion without necessarily solving for all constraint forces. The third approach will see the equations of motion derived using Lagrange's Equations. This is a work/energy approach that leads to the nonlinear differential equation of motion with minimal effort on the part of the analyst. The kinematic constraint forces are automatically eliminated as the governing equations are derived, leaving a designer with no information about forces acting on elements of the system unless extra work is done to find them. Lagrange's Equations are not typically introduced to undergraduate engineers as often as Newton's Laws are, so extra effort is made in this chapter to introduce the procedures for applying Lagrange's Equations to mechanical systems.

As an example, consider Figure 1.1. The figure shows a small bead with mass, m, sliding on a frictionless semicircular wire that rotates about a vertical axis with a constant angular velocity, ω. The wire has radius R. Gravity acts to pull the mass to the bottom of the semicircle while centripetal effects try to move it to the top. The single angular degree of freedom, θ, is sufficient to describe the motion of the bead on the wire.

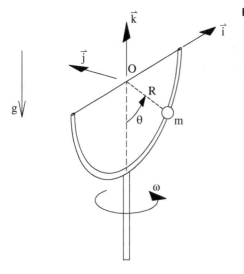

Figure 1.1 A bead on a wire.

1.1.1 Formal Vector Approach using Newton's Laws

Using the formal vector approach, the first step in the kinematic analysis is to choose a coordinate system (i.e. a set of unit vectors) that is convenient for expressing the vectors that will be used. The coordinate system may be fixed or rotating with some known angular velocity. In this case, we will use the $(\vec{\imath}, \vec{\jmath}, \vec{k})$ system shown in Figure 1.1. This is a rotating system fixed in the wire so that $\vec{\imath}$ and $\vec{k}$ stay in the plane of the wire and $\vec{\jmath}$ is perpendicular to the plane. Furthermore, $\vec{\imath}$ and $\vec{\jmath}$ remain horizontal and $\vec{k}$ is always vertical. The angular velocity of the coordinate system is $\vec{\omega} = \omega \vec{k}$.

We use the general approach to differentiating vectors, as follows, where $\vec{r}$ can be a position vector, a velocity vector, an angular momentum vector, or any other vector.

$$\frac{d\vec{r}}{dt} = \underbrace{\dot{\vec{r}}}_{\substack{\text{rate of change} \\ \text{of magnitude} \\ \text{of the vector}}} + \underbrace{\vec{\omega} \times \vec{r}}_{\substack{\text{rate of change} \\ \text{of direction} \\ \text{of the vector}}} \tag{1.1}$$

It is important to understand that the angular velocity vector, $\vec{\omega}$, is the absolute angular velocity of the coordinate system in which the vector, $\vec{r}$, is expressed. There is a danger that the rate of change of direction terms will be included twice if the angular velocity of the vector relative to the coordinate system in which it is measured is used instead.

We start the kinematic analysis by locating a fixed point, in this case point O, and writing an expression for the position vector that locates m with respect to O.

$$\vec{p}_{m/O} = R \sin \theta \vec{\imath} - R \cos \theta \vec{k} \tag{1.2}$$

The absolute velocity of m is

$$\vec{v}_m = \vec{v}_O + \frac{d}{dt}\vec{p}_{m/O} \tag{1.3}$$

Then, using Equation 1.1 and recognizing that $\vec{v}_O = 0$ since O is a fixed point and that $\dot{R} = 0$ since the radius of a semicircle is constant,

$$\vec{v}_m = (R\dot{\theta} \cos \theta \vec{\imath} + R\dot{\theta} \sin \theta \vec{k}) + \omega \vec{k} \times (R \sin \theta \vec{\imath} - R \cos \theta \vec{k}) \tag{1.4}$$

which can be simplified to

$$\vec{v}_m = R\dot{\theta} \cos \theta \vec{\imath} + \omega R \sin \theta \vec{\jmath} + R\dot{\theta} \sin \theta \vec{k} \tag{1.5}$$

The absolute acceleration of m is then

$$\vec{a}_m = \frac{d}{dt}\vec{v}_m = \dot{\vec{v}}_m + \omega \vec{k} \times \vec{v}_m \tag{1.6}$$

which simplifies to

$$\vec{a}_m = (R\ddot{\theta} \cos \theta - R\dot{\theta}^2 \sin \theta - R\omega^2 \sin \theta)\vec{\imath}$$
$$+ (2R\omega\dot{\theta} \cos \theta)\vec{\jmath} + (R\ddot{\theta} \sin \theta + R\dot{\theta}^2 \cos \theta)\vec{k} \tag{1.7}$$

Once an expression for the absolute acceleration has been found, the kinematic analysis is complete and we move on to drawing a Free Body Diagram (FBD). For this example, the FBD is shown in Figure 1.2.

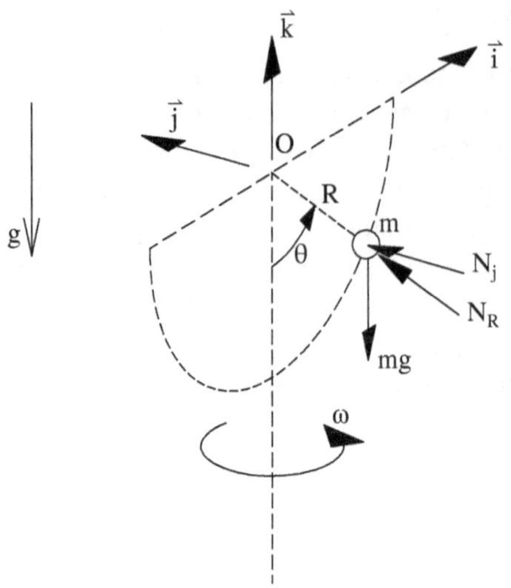

Figure 1.2 Free Body Diagram of a bead on a wire.

Constraints are taken into account when showing the forces acting on the bead. The forces shown and the rationale behind them are:

- $-mg\vec{k}$ = the weight of the body acting vertically downward. This is the effect of gravity.
- $N_j\vec{j}$ = one component of the normal force that the wire transmits to the mass. Since $\vec{j}$ is perpendicular to the plane of the wire, there can be a normal force in that direction.
- $\vec{N}_R = -N_R \sin\theta\vec{i} + N_R \cos\theta\vec{k}$ = the other component of the normal force. We let it have an unknown magnitude N_R and align it with the radial direction since that direction is normal to the wire.
- Note that there is no friction force because the system is frictionless. If there were, we would need to show a friction force acting in the direction that is tangential to the wire.

Once the FBD is complete, we can proceed to write Newton's Equations of Motion by simply summing forces in the positive coordinate directions and letting them equal the mass multiplied by the absolute acceleration in that direction. The result is three scalar equations as follows

$$\sum F_{\vec{i}} : -N_R \sin\theta = m(R\ddot{\theta}\cos\theta - R\dot{\theta}^2 \sin\theta - R\omega^2 \sin\theta) \tag{1.8}$$

$$\sum F_{\vec{j}} : N_j = m(2R\omega\dot{\theta}\cos\theta) \tag{1.9}$$

$$\sum F_{\vec{k}} : -mg + N_R \cos\theta = m(R\ddot{\theta}\sin\theta + R\dot{\theta}^2 \cos\theta) \tag{1.10}$$

At this point in the majority of undergraduate Dynamics courses we would count the number of unknowns that we have in the three equations to see if there is sufficient information to solve the problem. We would find five unknowns

$$N_R, N_j, \theta, \dot{\theta}, \ddot{\theta}$$

and say that we are unable to solve this without further information since we have only three equations. A typical textbook problem would say, for example, that the mass is released from rest (i.e. $\dot{\theta} = 0$) at a specified angle, θ_0, thereby removing two of the unknowns and letting you solve for N_R, N_j and $\ddot{\theta}$.

This solution gives an instantaneous look at the system that really doesn't point out the value of the equations derived. Equations 1.8 through 1.10 do not have five unknowns. They have two unknown constraint forces, N_R and N_j, and a group of variables $(\theta, \dot{\theta}, \ddot{\theta})$ that are related by differentiation. Rather than counting five unknowns as we did earlier, we should say that there are three unknowns

$$N_R, N_j, (\theta, \dot{\theta}, \ddot{\theta})$$

and three equations.

We can combine the three equations to eliminate N_R and N_j and we will be left with a single differential equation containing θ, $\dot{\theta}$, and $\ddot{\theta}$. This nonlinear, ordinary differential equation is the *equation of motion* for the system. Given initial conditions for θ and $\dot{\theta}$, we can solve the equation of motion as a function of time and predict the angle, its derivatives, and the two normal forces at any time. The solution of nonlinear differential equations is not a trivial exercise but can be handled fairly easily using numerical techniques.

The equation of motion for this system can be found by multiplying Equation 1.8 by $\cos\theta$ and adding the result to Equation 1.10 multiplied by $\sin\theta$, giving

$$mR\ddot{\theta} - mR\omega^2 \sin\theta\cos\theta + mg\sin\theta = 0 \tag{1.11}$$

Equation 1.9 is useful only for determining N_j during the motion. An expression for N_R can be found by multiplying Equation 1.8 by $\sin\theta$ and subtracting it from Equation 1.10 multiplied by $\cos\theta$. As a result, we could solve the differential equation of motion (Equation 1.11) numerically and always have the ability to predict the two constraint forces. These forces provide useful design information that is difficult to get from the methods considered next.

1.1.2 Informal Vector Approach using Newton's Laws

Here we consider a two-dimensional view of the system as shown in Figure 1.3 and work out the kinematic expressions for the accelerations from our knowledge of kinematics. There

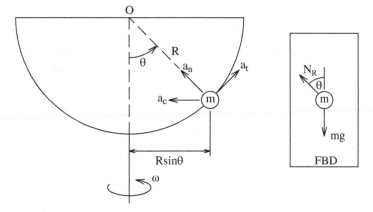

Figure 1.3 A 2D representation of the bead on a wire.

are three acceleration terms shown. They are a tangential acceleration, a_t, a normal acceleration, a_n, and a centripetal acceleration, a_c. Both a_t and a_n are due to the rates of change of the angle θ. Since the wire has constant radius, R, we can immediately write $a_t = R\ddot{\theta}$ and $a_n = R\dot{\theta}^2$ in the directions shown. Centripetal accelerations are of the form $a = \omega^2 r$ and a_c arises from the rotation of the wire with constant angular velocity ω. The relevant radius here is $R \sin \theta$ as shown. Therefore $a_c = \omega^2 R \sin\theta$ in the direction shown.

The inset in Figure 1.3 shows a FBD of the bead with the gravitational force and radial normal force being visible in this plane. There is another normal force perpendicular to the plane that can't be seen in this view. It is N_j in Figure 1.2 and was shown to be equal to $2mR\omega\dot{\theta} \cos \theta$ in Equation 1.9. The acceleration in this expression is a Coriolis acceleration. One needs quite a lot of experience with kinematic analysis to get the correct form of this term using an informal approach. Thankfully, it is perpendicular to the plane in which the bead moves relative to the wire, so it never appears in the equation of motion[1].

Summing forces in the vertical and horizontal directions gives

$$+ \uparrow \sum F : N_R \cos \theta - mg = ma_n \cos \theta + ma_t \sin \theta \tag{1.12}$$

$$+ \leftarrow \sum F : N_R \sin \theta = ma_c + ma_n \sin \theta - ma_t \cos \theta \tag{1.13}$$

To eliminate the constraining normal force from these two equations, we multiply Equation 1.12 by $\sin \theta$ and Equation 1.13 by $\cos \theta$ and subtract the resulting expressions. The result is

$$N_R (\cos \theta \sin \theta - \sin \theta \cos \theta) - mg \sin \theta =$$
$$ma_n (\cos \theta \sin \theta - \sin \theta \cos \theta) + ma_t (\sin^2\theta + \cos^2\theta) - ma_c \cos \theta \tag{1.14}$$

where it is clear that both N_R and ma_n are multiplied by zero and disappear from further consideration whereas ma_t is multiplied by a trigonometric identity equal to 1. Simplifying and substituting the derived kinematic expressions for a_t and a_c gives

$$mR\ddot{\theta} - mR\omega^2 \sin \theta \cos \theta + mg \sin \theta = 0 \tag{1.15}$$

which is the same nonlinear equation of motion (Equation 1.11) found in Subsection 1.1.1.

1.1.3 Lagrange's Equations of Motion

In this section, we consider the use of Lagrange's[2] Equations of Motion.

Lagrange's Equations, since they are based on work/energy principles, give the analyst two distinct advantages when deriving the equations of motion. First, the vector kinematic analysis is shorter than it is with a direct application of Newton's Laws since acceleration vectors need not be found. This is because the kinetic and potential energy expressions can

---

1 The normal force N_j is what determines the torque that some external mechanism must apply to the wire in order to enforce the constraint that it rotate with constant angular velocity. The analysis here, like many dynamic analyses, assumes that torque is available and simply works with the constraint on the angular velocity.

2 Joseph-Louis Lagrange (1736–1813), an Italian/French mathematician, is well known for his work on calculus of variations, dynamics, and fluid mechanics. In 1788 Lagrange published the *Mécanique Analytique* summarizing all the work done in the field of mechanics since the time of Newton, thereby transforming mechanics into a branch of mathematical analysis.

be derived from velocity vectors and position vectors respectively. Secondly, there is no need to draw free body diagrams for each of the rigid bodies in the system because the forces of constraint between the bodies do no work and are therefore not required for the analysis.

Of course, there are also disadvantages. The method requires a great deal of differentiation, sometimes of relatively complicated functions. Some analysts prefer the kinematics of Newton's method over the differentiation required when using Lagrange's Equations. Some point to a lack of physical feeling for problems without free body diagrams as being a disadvantage of the method. Finally, if the intent of analyzing the dynamics of a system is to predict loads, which could be carried forward into a structural analysis for instance, the forces of interaction between bodies are not available from a straightforward application of Lagrange's Equations.

1.1.3.1 The Bead on a Wire via Lagrange's Equations

We consider again the bead on the semicircular wire (Figure 1.1) and derive the equation of motion using Lagrange's Equations.

Lagrange's Equation is:

$$\frac{d}{dt}\left(\frac{\partial T}{\partial \dot{q}}\right) - \frac{\partial T}{\partial q} + \frac{\partial U}{\partial q} = Q_q \tag{1.16}$$

where:

- T = the total kinetic energy of the system
- U = the total potential energy of the system
- q = a generalized coordinate
- $\dot{q}$ = the time derivative of q
- Q_q = the generalized force corresponding to a variation of q

We first determine the kinetic energy of the system. This requires that we have an expression for the absolute velocity of the mass. This was done previously and the result, from Equation 1.5, is

$$\vec{v}_m = R\dot{\theta}\cos\theta\vec{i} + \omega R\sin\theta\vec{j} + R\dot{\theta}\sin\theta\vec{k} \tag{1.17}$$

The kinetic energy of the system is then

$$T = \frac{1}{2}m(\vec{v}_m \cdot \vec{v}_m) \tag{1.18}$$

which becomes, after substitution of Equation 1.17 and some simplification,

$$T = \frac{1}{2}mR^2(\dot{\theta}^2 + \omega^2\sin^2\theta) \tag{1.19}$$

Alternatively, using the informal approach and referring to Figure 1.3, we can see that there will be a component of velocity equal to $R\dot{\theta}$ tangent to the wire and another component equal to $\omega R\sin\theta$ perpendicular to the wire and into the page. These two components are mutually perpendicular so we can write, by applying Pythagoras' theorem,

$$T = \frac{1}{2}mv^2 = \frac{1}{2}m(R^2\dot{\theta}^2 + \omega^2R^2\sin^2\theta) \tag{1.20}$$

After factoring R^2 out of the brackets, this becomes exactly the same expression we had in Equation 1.19.

The potential energy of the system is due to gravity only. If the datum for potential energy is taken to be at point O, the potential energy, U, of the system is determined simply by the vertical distance from O to the bead. This distance is $R\cos\theta$ and, as the mass is below the datum, the potential energy is negative, leading to

$$U = -mgR\cos\theta \tag{1.21}$$

Having expressions for T and U and a single degree of freedom, θ, we can apply Lagrange's Equation (Equation 1.16) and find

$$q = \theta$$
$$\dot{q} = \dot{\theta}$$
$$\frac{\partial T}{\partial \dot{\theta}} = mR^2\dot{\theta}$$
$$\frac{d}{dt}\left(\frac{\partial T}{\partial \dot{\theta}}\right) = mR^2\ddot{\theta}$$
$$\frac{\partial T}{\partial \theta} = mR^2\omega^2\sin\theta\cos\theta$$
$$\frac{\partial U}{\partial \theta} = mgR\sin\theta$$
$$Q_\theta = 0 \tag{1.22}$$

Substituting the expressions from Equation 1.22 into Lagrange's Equation (Equation 1.16) gives the desired equation of motion

$$mR^2\ddot{\theta} - mR^2\omega^2\sin\theta\cos\theta + mgR\sin\theta = 0 \tag{1.23}$$

where we note that this equation when divided throughout by R yields the same result as Equations 1.11 and 1.15, the equation of motion derived using Newton's Laws.

Clearly, Equation 1.23 could be further simplified by factoring out the group mR but this would take away the ability to look at the individual terms and give a physical explanation for them. Whenever an equation is derived, the first test for correctness is to see if all of the terms have the same dimensions. In this case, the first term has dimensions of ML^2/T^2 where M is *mass*, L is *length*, and T is *time*. Note that angles such as θ are in radians, which are dimensionless since they are defined by an arc length divided by a radius. It follows that trigonometric functions such as $\sin\theta$ and $\cos\theta$ are also dimensionless. Angular velocities therefore have dimensions derived from angles divided by time, $1/T$, and angular acceler-ations are expressed as $1/T^2$. Using these conventions, it is easy to see that all three terms in Equation 1.23 have the same dimensions[3].

The dimensions of force are ML/T^2 or mass times acceleration. Taking this into account, we can see that the three terms in Equation 1.23 all have dimensions of FL or force times length. The terms are, in fact, all moments. The third term is the most obvious because it contains the gravity force mg multiplied by a moment arm of $R\sin\theta$. The moment arm

---

3 Note the difference between dimensions and units. *Dimensions* refer to physical characteristics such as mass, length, or time. *Units* refer to the system of measurement we use to substitute numbers into an equation. Examples are kilograms for mass, feet for length, and minutes for time.

is simply the horizontal distance between the mass and point O. Lagrange's Equation has produced an equation of motion based on a dynamic moment balance about the stationary point O and it did so without requiring the derivation of acceleration expressions, the drawing of free body diagrams, or the production of force and moment balance relationships. This is the power of using Lagrange's Equation for deriving equations of motion.

Following are explanations of terms that arise when using Lagrange's Equations.

1.1.3.2 Generalized Coordinates

The generalized coordinates are simply the degrees of freedom of the system with the condition that they be independently variable. That is, given a system with N generalized coordinates (q_i, $i = 1, N$), any generalized coordinate, q_j, must be able to undergo an arbitrary small variation, δq_j, with all of the other generalized coordinates being held constant.

In the example just considered, we might try to specify the position of the bead on the wire by using two coordinates – the vertical distance from O and the horizontal distance from O. We would soon find that these coordinates are not independent because the bead is constrained to stay on the circular wire so changing the horizontal position requires a change in the vertical position. These two coordinates are therefore not generalized coordinates.

1.1.3.3 Generalized Forces

The generalized force, Q_{q_r}, associated with the generalized coordinate, q_r, accounts for the effect of externally applied forces that are not included in the potential energy. We normally include elastic (i.e. spring) forces and gravitational forces in the potential energy and all others enter through the use of generalized forces.

Given a three-dimensional applied force

$$\vec{F} = F_x \vec{i} + F_y \vec{j} + F_z \vec{k} \tag{1.24}$$

with a position vector

$$\vec{p}_F = x\vec{i} + y\vec{j} + z\vec{k} \tag{1.25}$$

relative to a fixed point, we define the right-hand side of Lagrange's Equation for generalized coordinate q_r to be the *generalized force*, Q_{q_r}, where

$$Q_{q_r} = F_x \frac{\partial x}{\partial q_r} + F_y \frac{\partial y}{\partial q_r} + F_z \frac{\partial z}{\partial q_r} \tag{1.26}$$

The two most common methods for finding the generalized forces are as follows.

1. **The formal method**

   The most formal approach, and one that always works, starts with the vector expression of the absolute position of the point of application of the force

   $$\vec{p}_F = x\vec{i} + y\vec{j} + z\vec{k}$$

   and then writes the generalized force as

   $$Q_{q_r} = \vec{F} \cdot \frac{\partial \vec{p}_F}{\partial q_r} \tag{1.27}$$

Equation 1.27 can be written for each of N applied forces and the resulting scalar generalized forces can be added together to give the total generalized force for generalized coordinate q_r as

$$Q_{q_r} = \sum_{i=1}^{N} \vec{F}_i \cdot \frac{\partial \vec{p}_{F_i}}{\partial q_r} \tag{1.28}$$

2. **The intuitive approach**

Let there be n generalized coordinates specifying the position of a force acting on a dynamic system in Cartesian Coordinates. The force will be acting at the point (x, y, z) where the coordinates x, y and z are functions of the generalized coordinates q_1 through q_n and of time, t, as follows

$$x = x(q_1, q_2, \ldots, q_r, \ldots, q_n, t)$$
$$y = y(q_1, q_2, \ldots, q_r, \ldots, q_n, t)$$
$$z = z(q_1, q_2, \ldots, q_r, \ldots, q_n, t) \tag{1.29}$$

Variations in the position of the force as the generalized coordinates are varied while time is held constant can be written as

$$\delta x = \frac{\partial x}{\partial q_1}\delta q_1 + \frac{\partial x}{\partial q_2}\delta q_2 + \cdots \frac{\partial x}{\partial q_r}\delta q_r + \cdots \frac{\partial x}{\partial q_n}\delta q_n$$
$$\delta y = \frac{\partial y}{\partial q_1}\delta q_1 + \frac{\partial y}{\partial q_2}\delta q_2 + \cdots \frac{\partial y}{\partial q_r}\delta q_r + \cdots \frac{\partial y}{\partial q_n}\delta q_n$$
$$\delta z = \frac{\partial z}{\partial q_1}\delta q_1 + \frac{\partial z}{\partial q_2}\delta q_2 + \cdots \frac{\partial z}{\partial q_r}\delta q_r + \cdots \frac{\partial z}{\partial q_n}\delta q_n \tag{1.30}$$

If we are trying to find the generalized force corresponding to only one of the generalized coordinates, say q_r, we rewrite Equation 1.30 with $\delta q_i = 0; i = 1, n; i \neq r$ and $\delta q_r \neq 0$, giving

$$\delta x = \frac{\partial x}{\partial q_r}\delta q_r$$
$$\delta y = \frac{\partial y}{\partial q_r}\delta q_r$$
$$\delta z = \frac{\partial z}{\partial q_r}\delta q_r \tag{1.31}$$

Now consider Equation 1.26 with each side multiplied by δq_r

$$Q_{q_r}\delta q_r = F_x\frac{\partial x}{\partial q_r}\delta q_r + F_y\frac{\partial y}{\partial q_r}\delta q_r + F_z\frac{\partial z}{\partial q_r}\delta q_r \tag{1.32}$$

The terms from Equation 1.31 can be substituted into the right-hand side of Equation 1.32 to yield

$$Q_{q_r}\delta q_r = F_x\delta x + F_y\delta y + F_z\delta z \tag{1.33}$$

The right-hand side of Equation 1.33 can be seen to be the work done by the applied force as its position varies due to changes in the generalized coordinate q_r while all other generalized coordinates and time are held constant.

Using the intuitive approach to finding generalized forces, the analyst will consider, in sequence, the variation of individual generalized coordinates and will write expressions for the total work done during each variation. The generalized force associated with each generalized coordinate will be the work done during the variation of that coordinate, δW_{q_r}, divided by the variation in the coordinate. That is,

$$Q_{q_r} = \delta W_{q_r} / \delta q_r \tag{1.34}$$

1.1.3.4 Dampers – Rayleigh's Dissipation Function

Devices called "dampers" are common in mechanical systems. These are elements that dissipate energy and they are modeled as producing forces that are proportional to their rate of change of length. The rate of change of length is the relative velocity across the damper. "Proportional" implies linearity and a force proportional to speed implies laminar, viscous flow. As a result, these elements are often referred to as "linear viscous dampers".

Figure 1.4 shows a system where a body is attached to ground by a damper. The body is moving to the right with speed v and the damping coefficient (constant of proportionality) is c. The physical connection of the damper to both the ground and the body dictates that the rate of change of length of the damper is equal to the speed v. The force in the damper will therefore be $F_d = cv$. The direction of the force will be such that it causes the damper to increase in length as shown in the lower part of Figure 1.4. By Newton's 3rd Law, the force on the body must be equal and opposite to the force acting on the damper. The force F_d therefore acts to the left on the body. In other words, the damping force opposes the velocity of the body.

Consider now the more general case of a particle where the velocity of the body is given by

$$\vec{v}_m = \dot{x}\vec{i} + \dot{y}\vec{j} + \dot{z}\vec{k} \tag{1.35}$$

Given this velocity, the force that the damper applies to the particle will be

$$\vec{F}_d = -c\vec{v}_m = -c\dot{x}\vec{i} - c\dot{y}\vec{j} - c\dot{z}\vec{k} \tag{1.36}$$

The components of $\vec{F}_d$ can be substituted into Equation 1.26 to get the following expression for the generalized force arising from the damper

$$Q_{q_r}^d = -c\dot{x}\frac{\partial x}{\partial q_r} - c\dot{y}\frac{\partial y}{\partial q_r} - c\dot{z}\frac{\partial z}{\partial q_r} \tag{1.37}$$

Figure 1.4 A linear viscous damper.

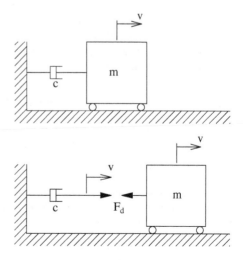

where we can write[4]

$$\frac{\partial x}{\partial q_r} = \frac{\partial \dot{x}}{\partial \dot{q}_r}; \frac{\partial y}{\partial q_r} = \frac{\partial \dot{y}}{\partial \dot{q}_r}; \frac{\partial z}{\partial q_r} = \frac{\partial \dot{z}}{\partial \dot{q}_r} \tag{1.38}$$

which can be substituted into Equation 1.37 to yield the following expression for the generalized force

$$Q_{q_r}^d = -c\left(\dot{x}\frac{\partial \dot{x}}{\partial \dot{q}_r} + \dot{y}\frac{\partial \dot{y}}{\partial \dot{q}_r} + \dot{z}\frac{\partial \dot{z}}{\partial \dot{q}_r} \right) \tag{1.39}$$

The generalized force, as expressed in Equation 1.39, can be derived from a scalar function called *Rayleigh's Dissipation Function* which is defined as

$$\Re = \frac{1}{2}\, c(\vec{v}_m \cdot \vec{v}_m) = \frac{1}{2}c(\dot{x}^2 + \dot{y}^2 + \dot{z}^2) \tag{1.40}$$

A simple differentiation with respect to $\dot{q}_r$ yields

$$\frac{\partial \Re}{\partial \dot{q}_r} = c\left(\dot{x}\frac{\partial \dot{x}}{\partial \dot{q}_r} + \dot{y}\frac{\partial \dot{y}}{\partial \dot{q}_r} + \dot{z}\frac{\partial \dot{z}}{\partial \dot{q}_r} \right) = -Q_{q_r}^d \tag{1.41}$$

Lagrange's Equation can then be written as

$$\frac{d}{dt}\left(\frac{\partial T}{\partial \dot{q}_r} \right) - \frac{\partial T}{\partial q_r} + \frac{\partial U}{\partial q_r} = -\frac{\partial \Re}{\partial \dot{q}_r} + Q_{q_r} \tag{1.42}$$

where Q_{q_r} now represents the generalized force corresponding to all externally applied forces that are neither conservative nor linear viscous in nature. Finally, we can transfer the Rayleigh Dissipation term to the left-hand side and write Lagrange's Equation with dissipation as

$$\frac{d}{dt}\left(\frac{\partial T}{\partial \dot{q}_r} \right) - \frac{\partial T}{\partial q_r} + \frac{\partial U}{\partial q_r} + \frac{\partial \Re}{\partial \dot{q}_r} = Q_{q_r} \tag{1.43}$$

1.2 Equilibrium Solutions

Equilibrium solutions of the equations of motion are those where the degrees of freedom assume values which cause their first and second derivatives to go to zero. Under these conditions, there will be no tendency for the values of the degrees of freedom to change and the system will be in an *equilibrium state*.

Some equilibrium states are stable and some are unstable and, inevitably, *systems are either in a stable equilibrium state or trying to get to one*. The study of small motions around a stable equilibrium state is called *Vibrations*.

1.2.1 Equilibrium of a Simple Pendulum

We start by considering the simple pendulum[5] shown in Figure 1.5. Using the angle θ as the single degree of freedom, the equation of motion is

$$m\ell^2\ddot{\theta} + mg\ell \sin\theta = 0 \tag{1.44}$$

---

4 To see that this is the case, first divide both sides of Equation 1.30 by δt, then let δt go to zero and note that the left-hand sides become $\dot{x}, \dot{y}$, and $\dot{z}$ respectively. On the right-hand side, $\delta q_i/\delta t$ becomes $\dot{q}_i$ for all i. Finally, take the partial derivatives with respect to $\dot{q}_r$ and the expressions in Equation 1.38 result.
5 The definition of the often-quoted *simple pendulum* is that it has a massless rigid rod supporting a point mass. The rod is free to swing in a plane about the frictionless point where it is connected to the ground. The only external force is that due to gravity.

Figure 1.5 A simple pendulum.

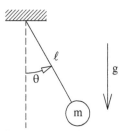

Once started in motion the pendulum will swing about the point of connection to the ground. In the case of the simple pendulum there is no mechanism for removing energy from the system as it swings (i.e. no friction or other forces that do work) so the motion, once started, will persist.

The motion will depend on the way in which it is started. That is, if the pendulum is rotated to some arbitrary starting angle, θ_0, and released from rest, it will swing through the position where $\theta = 0$ and will eventually return to where it started before reversing and starting the cyclic motion over again. If the pendulum is stopped and returned to θ_0 and then released, not from rest but with an initial velocity, the resulting motion will be different and the pendulum will pass through θ_0 when it returns. The motion will, however, still be cyclic.

The question we ask now is *Are there initial values of θ where the pendulum can be released from rest and remain stationary?* These are the equilibrium states.

Consider Equation 1.44 under the conditions that there is an initial angle θ_0 and there is no angular velocity (i.e. $\dot{\theta} = 0$ so that θ_0 does not change with time) and, further, that there is no angular acceleration (i.e. $\ddot{\theta} = 0$ so that $\dot{\theta}$ does not change with time and thus there will never be a change in θ_0). This is an equilibrium position and Equation 1.44 becomes the *equilibrium condition.*

$$mg\ell \sin \theta_0 = 0 \tag{1.45}$$

Since m, g, and ℓ are never zero, this can only be satisfied by:

$$\sin \theta_0 = 0$$

The total range of θ is $0 \leq \theta \leq 2\pi$. In this range, only $\theta_0 = 0$ (the pendulum hangs vertically downward) and $\theta_0 = \pi$ (the pendulum stands upright) satisfy the requirements. These are the two equilibrium states for the pendulum.

There are formal methods for testing the stability of the equilibrium states but that we leave to courses on control systems. It is sufficient for us to be able to see that the state where the pendulum stands upright is unstable and the pendulum will try to get to the stable equilibrium position where $\theta_0 = 0$.

The vibrations question is *What will be the response of the system for small motions away from the stable equilibrium condition where $\theta_0 = 0$?*

1.2.2 Equilibrium of the Bead on the Wire

We now return to our continuing example problem – the bead on a rotating semicircular wire as shown in Figure 1.1. The equation of motion (see Equation 1.23) is

$$mR^2\ddot{\theta} - mR^2\omega^2 \sin \theta \cos \theta + mgR \sin \theta = 0 \tag{1.46}$$

We look for solutions where the angle θ remains constant. To find these solutions, we let $\theta = \theta_0$ and set $\dot{\theta} = \ddot{\theta} = 0$ so that the angle can never change. This results in the *equilibrium condition*

$$-mR^2\omega^2 \sin\theta_0 \cos\theta_0 + mgR \sin\theta_0 = 0 \tag{1.47}$$

The equilibrium condition is a group of constant terms summing up to zero that becomes an identity for us. We will see this group of terms again when we write the equation of motion for small motions around equilibrium and every time we see it, we will be able to set it equal to zero.

With some simple factoring out of terms, we get

$$mR \sin\theta_0 (-R\omega^2 \cos\theta_0 + g) = 0 \tag{1.48}$$

This expression will hold for two cases:

- $\sin\theta_0 = 0$. This is satisfied when $\theta_0 = 0$ and when $\theta_0 = \pi$. These correspond to the bead being directly below point O and directly above point O respectively. Being above point O is, of course, physically impossible for the semicircular wire but would be possible for a complete hoop.
- $(-R\omega^2 \cos\theta_0 + g) = 0$. This is satisfied if

$$\cos\theta_0 = \frac{g}{R\omega^2}$$

This is an equilibrium value of θ where the gravitational pull and the centripetal effects exactly balance each other. It corresponds to an angle between $-\pi/2$ and $\pi/2$ because the positive values of g, R, and ω^2 force the cosine to be positive. We will be interested in the behavior of the bead for small motions about this equilibrium state.[6]

1.3 Linearization

There two types of nonlinearities that we will often be required to deal with. They are (1) geometric nonlinearities and (2) structural nonlinearities. Geometric nonlinearities arise from trigonometric functions of large angles and structural nonlinearities are due to the inherent nonlinear stiffness (i.e. force versus deflection characteristic) of materials for large deflections.

1.3.1 Geometric Nonlinearities

Both the simple pendulum with the EOM (Equation of Motion)

$$m\ell^2\ddot{\theta} + mg\ell \sin\theta = 0 \tag{1.49}$$

and the bead on a rotating wire with the EOM

$$mR^2\ddot{\theta} - mR^2\omega^2 \sin\theta \cos\theta + mgR \sin\theta = 0 \tag{1.50}$$

---

6 All of these equilibrium states have very interesting stability characteristics but considering stability is outside the scope of what we are doing here. It is sufficient for our purposes to say, without proof, that the equilibrium state between $-\pi/2$ and $\pi/2$ is stable.

have geometric nonlinearities arising from the sine and cosine terms. They will both be addressed in the following, starting with the simple pendulum.

1.3.1.1 Linear EOM for a Simple Pendulum

The nonlinear differential equation of motion for the pendulum (Equation 1.44) is valid for any range of motion. The difficulty is that we can't solve nonlinear differential equations without resorting to numerical methods. We get around this problem by *linearizing* the differential equation because we have had courses on how to solve linear differential equations.

To do this, we consider small motions near the stable equilibrium state. Let θ in Equation 1.44 be replaced by $\theta_0 + \beta$ where β is a very small angle and θ_0 is the equilibrium value of θ. That is,

$$\theta = \theta_0 + \beta$$

and, differentiating with the knowledge that θ_0 is constant, we find

$$\dot{\theta} = \dot{\theta}_0 + \dot{\beta} = \dot{\beta}$$

and then

$$\ddot{\theta} = \ddot{\beta}$$

Substituting into Equation 1.44 yields

$$m\ell^2\ddot{\beta} + mg\ell \sin(\theta_0 + \beta) = 0 \tag{1.51}$$

We can use the trigonometric identity for the sine of the sum of two angles[7] to write

$$\sin(\theta_0 + \beta) = \cos\theta_0 \sin\beta + \sin\theta_0 \cos\beta \tag{1.52}$$

---

7 While the "sum of angles formulae" may be "well known", they are not easily remembered. However, they can be quickly derived using Euler's equation

$$e^{i\theta} = \cos\theta + i\sin\theta$$

We write the equation twice, once for the angle α and again for the angle β, giving

$$e^{i\alpha} = \cos\alpha + i\sin\alpha$$

$$e^{i\beta} = \cos\beta + i\sin\beta$$

We then multiply these two equations together giving, on the left-hand side,

$$e^{i\alpha}e^{i\beta} = e^{i(\alpha+\beta)} = \cos(\alpha + \beta) + i\sin(\alpha + \beta)$$

and, on the right-hand side,

$$(\cos\alpha \cos\beta - \sin\alpha \sin\beta) + i(\cos\alpha \sin\beta + \sin\alpha \cos\beta)$$

Equating the real and imaginary parts of these two results gives the "sum of angles" formulae we couldn't remember

$$\cos(\alpha + \beta) = \cos\alpha \cos\beta - \sin\alpha \sin\beta$$

$$\sin(\alpha + \beta) = \cos\alpha \sin\beta + \sin\alpha \cos\beta$$

We now consider rewriting Equation 1.52 under the condition where β is a very small angle. The small angle conditions on the sine and cosine are derived from their Maclaurin's Series expansions. The expansions are

$$\sin \beta = \beta - \frac{\beta^3}{3!} + \frac{\beta^5}{5!} - \frac{\beta^7}{7!} + \cdots$$

and

$$\cos \beta = 1 - \frac{\beta^2}{2!} + \frac{\beta^4}{4!} - \frac{\beta^6}{6!} + \cdots$$

If β is very small, then higher powers of β are much smaller and are negligible[8] in the series. We therefore approximate the sine and cosine with

$$\sin \beta \approx \beta$$

and

$$\cos \beta \approx 1$$

Equation 1.52 can then be written as

$$\sin(\theta_0 + \beta) = \cos \theta_0 \beta + \sin \theta_0 \tag{1.53}$$

and the equation of motion (Equation 1.51) becomes

$$m\ell^2 \ddot{\beta} + mg\ell \cos \theta_0 \beta + mg\ell \sin \theta_0 = 0 \tag{1.54}$$

Note the constant term $mg\ell \sin \theta_0$ that appears in Equation 1.54. This is the term that was set to zero to determine the equilibrium state (see Equation 1.45) and it is still equal to zero so it can be removed. The equilibrium condition is always a set of constant terms that must be zero for the system not to move and that set of constant terms always reappears in the linearized equation of motion. After removing the equilibrium condition, the linearized equation of motion for the pendulum becomes

$$m\ell^2 \ddot{\beta} + mg\ell \cos \theta_0 \beta = 0 \tag{1.55}$$

This equation is valid for motion about either of the two equilibrium states we found (i.e. $\theta_0 = 0$ with $\cos \theta_0 = 1$ and $\theta_0 = \pi$ with $\cos \theta_0 = -1$). We write

$$m\ell^2 \ddot{\beta} + mg\ell \beta = 0 \text{ for } \theta_0 = 0 \tag{1.56}$$

and

$$m\ell^2 \ddot{\beta} - mg\ell \beta = 0 \text{ for } \theta_0 = \pi \tag{1.57}$$

Equation 1.56 will yield oscillating solutions (i.e. vibrations – more to come later) and Equation 1.57 will yield growing exponential solutions, showing that the system really doesn't want to stay in the unstable upright position.

------

8 In the case of an angle of ten degrees, for example, we convert the angle to 0.174533 radians. The percentage errors using the linear approximations are only 0.5% on the sine and 1.5% on the cosine. Ten degrees of rotation is a very large angle in the world of vibrations. We are typically looking at fractions of a degree where the linear approximations are very accurate.

1.3.1.2 Linear EOM for the Bead on the Wire

The EOM for the bead on the rotating wire is

$$mR^2\ddot{\theta} - mR^2\omega^2 \sin\theta \cos\theta + mgR \sin\theta = 0 \qquad (1.58)$$

As we did for the simple pendulum, we let $\theta = \theta_0 + \beta$ where β is a small angle. Since θ_0 is a constant, we differentiate to find that $\dot{\theta} = \dot{\beta}$ and $\ddot{\theta} = \ddot{\beta}$. Substituting into Equation 1.58 gives

$$mR^2\ddot{\beta} - mR^2\omega^2 \sin(\theta_0 + \beta)\cos(\theta_0 + \beta) + mgR \sin(\theta_0 + \beta) = 0 \qquad (1.59)$$

We use the trigonometric identities

$$\cos(\theta_0 + \beta) = \cos\theta_0 \cos\beta - \sin\theta_0 \sin\beta$$

$$\sin(\theta_0 + \beta) = \cos\theta_0 \sin\beta + \sin\theta_0 \cos\beta$$

along with the linearizing approximations for the small angle β

$$\sin\beta \approx \beta$$

$$\cos\beta \approx 1$$

to get the linearized forms

$$\cos(\theta_0 + \beta) \approx \cos\theta_0 - (\sin\theta_0)\beta$$

$$\sin(\theta_0 + \beta) \approx (\cos\theta_0)\beta + \sin\theta_0$$

Then, the term $\sin(\theta_0 + \beta)\cos(\theta_0 + \beta)$ which appears in Equation 1.59 can be approximated by the product of these two expressions

$$\sin(\theta_0 + \beta)\cos(\theta_0 + \beta) = [\cos\theta_0 - (\sin\theta_0)\beta][(\cos\theta_0)\beta + \sin\theta_0]$$
$$= -\cos\theta_0 \sin\theta_0 \beta^2 + (\cos^2\theta_0 - \sin^2\theta_0)\beta + \cos\theta_0 \sin\theta_0 \quad (1.60)$$

We then say that the nonlinear β^2 term can be neglected as being negligibly small compared to the linear term β since β is a small angle. This is at the heart of the linearization process. The linearized EOM becomes

$$mR^2\ddot{\beta} - mR^2\omega^2 (\cos^2\theta_0 - \sin^2\theta_0)\beta - mR^2\omega^2 \cos\theta_0 \sin\theta_0$$
$$+ mgR(\cos\theta_0\beta + \sin\theta_0) = 0 \qquad (1.61)$$

Now we rearrange this to separate the constant terms from the terms with β and its derivatives. The result is

$$mR^2\ddot{\beta} + [-mR^2\omega^2 (\cos^2\theta_0 - \sin^2\theta_0) + mgR \cos\theta_0]\beta$$
$$+ [-mR^2\omega^2 \sin\theta_0 \cos\theta_0 + mgR \sin\theta_0] = 0 \qquad (1.62)$$

where the constant term $[-mR^2\omega^2 \sin\theta_0 \cos\theta_0 + mgR \sin\theta_0]$ is exactly the same as the equilibrium condition stated in Equation 1.47 and it is identically equal to zero by definition so it can be removed from the equation of motion. This will always be the case for vibrational systems. As we progress through the following chapters, we will often say something to the effect of "the gravitational forces are canceled out by preloads in the springs so they can be ignored" where we actually mean that our equations of motion are written for small motions away from an equilibrium state and that we need only be concerned with how the forces in elements change as the system moves slightly away from equilibrium. There will be forces holding the system in equilibrium and they may be large but they always add up to zero. The equilibrium condition for the bead is actually a statement that the moment that the gravitational force exerts about point O is exactly equal and opposite to the moment about O due to the centripetal acceleration.

It remains to pick a suitable equilibrium state from the three we found in Subsection 1.2.2. Of these, the most interesting is the one where the bead is not below or above point O. That is, consider the case where

$$\cos \theta_0 = \frac{g}{R\omega^2}$$

With a little effort and use of one trigonometric identity, you can show that the linear equation of motion about this state can be written as

$$\ddot{\beta} + \left[\frac{\omega^4 - (g^2/R^2)}{\omega^2} \right] \beta = 0 \tag{1.63}$$

As you will see later, this is the standard form for an undamped vibrational equation of motion. If the bead is in equilibrium and is disturbed away from equilibrium by a small angle, it will begin to oscillate with a frequency that can be determined directly from Equation 1.63.[9]

1.3.2 Nonlinear Structural Elements

Making linear approximations to trigonometric functions is not the only consideration we have in creating linear differential equations of motion. Many times the physical properties of structural elements in the system are nonlinear. A rubber suspension element is a good example. Depending on how it is designed, it can be made to get softer or harder as it deflects. Note that "softer" and "harder" are non-technical words relating to the *stiffness* of the element. Figure 1.6 shows the characteristics of a "hardening" spring where the element gets stiffer as it deflects. The stiffness is measured by the local tangent to the curve.

The equilibrium solution to the nonlinear equation of motion will place the system in equilibrium at x_0 (the point labeled *operating point* on Figure 1.6). Once there, we consider motions Δx away from the operating point and use a Taylor's Series expansion for the nonlinear function $f(x)$

$$f(x_0 + \Delta x) = f(x_0) + \frac{df}{dx}\bigg|_{x=x_0} \Delta x + \frac{1}{2!}\frac{d^2f}{dx^2}\bigg|_{x=x_0} \Delta x^2 + \frac{1}{3!}\frac{d^3f}{dx^3}\bigg|_{x=x_0} \Delta x^3 + \cdots \tag{1.64}$$

For small values of Δx, we can neglect the higher order terms and write

$$f(x_0 + \Delta x) \approx f(x_0) + \frac{df}{dx}\bigg|_{x=x_0} \Delta x \tag{1.65}$$

---

9 For those readers who already know about the standard form of these equations and are worried about the consequences of the term in square brackets being negative, I suggest you take time to consider what value $\cos \theta_0$ must have for this term to be negative.

Figure 1.6 Nonlinear structural
element – Linearization and effective
stiffness.

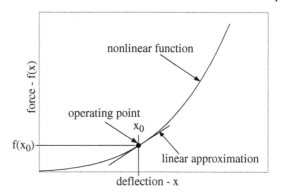

where we can see that the linear term involves only the first derivative of the function. This
derivative is the local tangent to the curve and is the *effective stiffness* of the element for this
operating point. Linearization of functions about operating points is therefore an exercise
in finding the local slope of the function and assuming that small deviations away from
the operating point can be approximated by points lying on this straight line. The linear
approximation is shown graphically in Figure 1.6.

The constant value of the force at the operating point $f(x_0)$ is the force acting at equilibrium and will enter the equation of motion in such a way that it and constant forces in other
elements in the system will sum to zero. We consider these constant forces to be *preloads* on
the elements and will quickly fall into the habit of leaving them out of the analysis because
they always add up to zero.

1.4 Summary

The study of *Vibrations* is the study of the behavior of dynamic systems as they experience small motions around stable equilibrium states. The steps in arriving at the governing
equation[10] of motion are

- The nonlinear equations of motion for a system can be derived using either Newton's
  Laws or Lagrange's Equation.
- The nonlinear equations of motion can be used to find equilibrium states for the system.
- The degrees of freedom of the system can be replaced by their stable equilibrium values
  plus very small variables representing motions away from equilibrium.
- The equations of motion can be linearized so that small motions about the stable equi-
  librium state are governed by linear, ordinary, differential equations.

Exercises

1.1 The figure shows an undeflected, nonlinear spring on the left and the same spring,
now deflected by the weight of the hanging mass, on the right. Taking the coordinate,

---

10 All of the analysis here has assumed that the motion of the system can be described with a single degree
of freedom and therefore a single equation of motion. We will soon see that this is not the case but that the
material and methods presented here still apply.

x, as the deflection of the spring from its free length, the force in the spring can be expressed as

$$F_s = (10^7)x^2 + (10^5)x$$

where F_s is in Newtons and x is in meters.

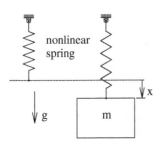

Figure E1.1

nonlinear
spring

(a) What will be the equilibrium value of x for a mass of 1200 kg?
(b) What is the effective stiffness of the system for small motions around the equilibrium in part (a)?
(c) Plot the force versus deflection curve for the element. Show the operating point and the linear approximation to be used in the equation of motion.

1.2 The figure shows the force/deflection characteristic for hardening springs used to support a large machine in a factory. The force (F in Newtons) written as a function of deflection (x in mm) is

$$F = 100x^3 + 1000x$$

Four mounts of this type are placed under the four corners of the machine and deflect 3 mm as they take up the weight.

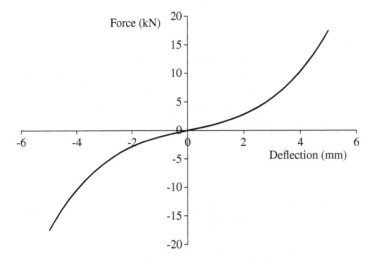

Figure E1.2

(a) What effective stiffness does the system have about equilibrium?

(b) What is the mass of the machine?

1.3 The effective stiffness of any object that deflects under the action of a force can be expressed as the ratio of that force divided by the resulting deflection. In some cases, there are good theoretical expressions that can be used to get the effective stiffness. A good example is the effective stiffness of a uniform beam. Suppose you have two identical lightweight, uniform beams of length L and flexural rigidity EI. One of the beams is cantilevered and the other is simply supported. You have a choice of supporting a heavy piece of equipment at the end of the cantilevered beam or at the center of the simply supported beam. Which support has a greater effective stiffness? What is the ratio of the two effective stiffnesses?

1.4 An open cylindrical container with a weight at its bottom is placed in the ocean. The cylinder sinks until buoyancy forces equal the total weight and then floats upright in equilibrium.

(a) Draw a FBD of the cylinder just after it is placed in the water. Use a DOF, y, to indicate the distance the bottom of the cylinder has traveled below the free surface of the water. There will be a buoyancy force acting upward on the cylinder. This force is due to the water pressure a distance y under the surface. That pressure is equal to $\rho g y$ where ρ is the density of water and g is the acceleration due to gravity. Assume a cross-sectional area of A for the cylinder.

(b) Use Newton's Laws to write the equation of motion for the cylinder.

(c) Find the equilibrium condition for the cylinder and solve for y_0, the equilibrium value of y.

1.5 The figure shows a cart of mass m that is attached to the ground at point A by a spring of stiffness k and a damper with damping coefficient c. The motion of the cart is forced by a harmonic motion $y(t) = Y \sin \omega t$ at the end of another damper with coefficient c.

Write the equation of motion for the system using $x(t)$ as the degree of freedom. Do this once using Newton's Laws and again using Lagrange's Equation and confirm that you get the same result from each method.

Figure E1.5

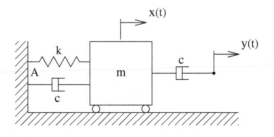

1.6 The figure shows a thin, uniform rod with mass $= m$ and length $= \ell$ that has been released from rest where $\theta = 0$. The rod rotates about a frictionless pin at A.

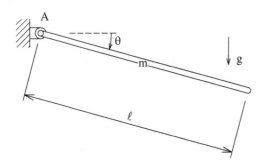

Figure E1.6

(a) Write the nonlinear equation of motion for the rod using θ as the DOF.
(b) Derive the equilibrium condition.
(c) What are the two equilibrium values of θ?
(d) Linearize the equation of motion about each of the equilibrium states and compare the two differential equations. The result should be two, linear, second-order, homogeneous differential equations that differ only in the sign of the coefficient multiplying θ. This coefficient is the "stiffness" and having a negative stiffness makes an equilibrium state unstable. Does your unstable state make physical sense?

2

Single Degree of Freedom Systems – Modeling

We move on to considering the response of single degree of freedom (SDOF) systems for small motions about stable equilibrium states. These systems are governed by linear, ordinary, differential equations of motion and we consider their response to initial conditions and to harmonic applied forces of various forms, with and without damping.

Understanding the response of SDOF systems is crucial to understanding vibrations. Even when we move on to systems with many degrees of freedom (e.g. a finite element model of a complex structure with 10,000 degrees of freedom) we will see that the response of the systems can be described using the same characteristics developed here for SDOF systems.

2.1 Modeling Single Degree of Freedom Systems

We start with a simple mass on a spring which is the archetypal model of a single degree of freedom system. Figure 2.1 shows the model where the block labeled m indicates that there is mass in the system and the element labeled k indicates that the mass is supported by a structure that has an effective stiffness of k. The single degree of freedom, x, is defined to be the vertical motion of the mass. It is important to note that x is defined to be positive upward by virtue of the fact that the arrowhead shows it that way. Once we select a positive direction for a degree of freedom we must respect it as we do the kinematics and the force balances. Signs will be lost in the equation of motion if the direction is not used consistently.

Before we go too far with the spring/mass model we should understand that structures we are dealing with are never as simple as a block of mass resting on a single spring. We are more likely to be dealing with a structure like the independent front suspension of an automobile shown in Figure 2.2. This structure has a single degree of freedom because the movement of all of the components of the four bar linkage making up the structure can be completely specified if we know the motion of any point in the system. The inclined coil spring connecting the lower control arm to the chassis provides the system stiffness. It deflects as the tire moves relative to the chassis and, for the small motions we consider in vibrations, the resulting force in the spring will be proportional to the tire movement thereby giving the system an effective stiffness. The wheel/tire assembly and the links in the suspension have mass and the effective mass of the system can be worked out from them and the system kinematics. Taking the chassis as ground in this system allows the

Introduction to Mechanical Vibrations, First Edition. Ronald J. Anderson.
© 2020 John Wiley & Sons Ltd. Published 2020 by John Wiley & Sons Ltd.
Companion website: www.wiley.com/go/anderson/introduction-to-vibrations

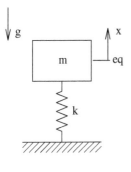

Figure 2.1   A mass on a spring.

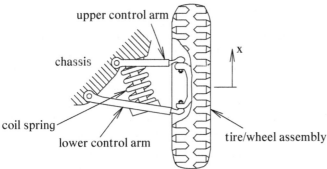

Figure 2.2   Independent front suspension.

motion of every element to be completely specified by the vertical motion of the tire/wheel assembly, so we can work with the single degree of freedom x.

Going back to Figure 2.1, you will see a horizontal line labeled "eq" with the arrow indicating the degree of freedom, x, extending from it. This simply indicates that x is defined as a small motion away from the stable equilibrium position. The "eq" line is fixed in space and can be used as an inertial reference point for kinematics. Thus the absolute velocity and acceleration of the mass are $\dot{x}$ and $\ddot{x}$ respectively, both being positive[1] in the same direction as positive x.

2.1.1   Deriving the Equation of Motion

We consider first the spring alone as shown on the left in Figure 2.3. There is no force acting on the spring so it is undeflected or at its "free length". The mass is then added to the spring and we define the degree of freedom, y, as shown on the right side of the figure.

Consider deriving the equation of motion using Newton's Laws. The kinematic analysis is very simple, with the absolute acceleration being $\ddot{y}$ downwards since y is measured

---

1 The positive directions for the velocity and acceleration come strictly from the definition of derivatives of functions. If a function is increasing in a given direction, then its slope or derivative is positive in that direction. It is very easy to lose a sign when deriving the equation of motion if you try to bring physical reasoning into the derivation. Physically, if the mass is above the equilibrium point, then the spring is trying to pull it back to equilibrium and the acceleration will be downward. Using this in the analysis will cause the loss of a sign because, while being true, it simply indicates that the acceleration is negative with respect to the chosen coordinate system when the mass is above equilibrium.

Figure 2.3 Assembly of the mass/spring system.

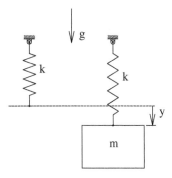

Figure 2.4 FBD of the mass/spring system.

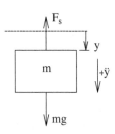

with respect to the fixed horizontal line passing through the end of the undeflected spring. Figure 2.4 shows the free body diagram of the mass. The mass is assumed to have moved in the positive y-direction. The gravitational force, mg, and the spring force, F_s, are shown. Applying Newton's 2nd Law yields

$$+\downarrow \sum F : mg - F_s = m\ddot{y} \tag{2.1}$$

At this point we need to consider the form of the spring force.

When modeling, every element in a system is characterized by two variables and the way in which those variables are related to each other. Given an element, there is always a variable that passes through it and another that is measured across it. Forces pass through springs and displacements are measured across springs; currents flow through resistors and voltage is measured across resistors; and so on. The force in a linear spring is proportional to the displacement (Hooke's Law) and the current in a resistor is proportional to the voltage (Ohm's Law). These two laws are called the constitutive relationships for the spring and the resistor respectively. Every element has a constitutive relationship that says how the variable passing through it is related to the variable measured across it.

A spring element and its variables are shown in Figure 2.5. The force acting through the spring, F_s, must be equal and opposite at the ends of the spring in order to keep it in equilibrium[2]. In this case, we have chosen to define the force as if the spring is in tension. The motions of the spring ends are defined as x_1 on the left and x_2 on the right where x_1 and x_2 are motions dependent on the DOF choices we have made in modeling the system.

---

2 Notice that the spring has been assumed to be massless in order to make this statement. This is a common assumption for the *lumped-parameter* models typically used in vibrations. Mass elements are assumed to be rigid so they provide no elastic forces and spring elements are assumed to be massless so that their accelerations do not enter the force balance equations. How to deal with the mass of springs is discussed in Section 12.1

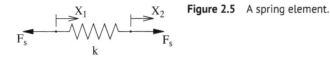

Figure 2.5 A spring element.

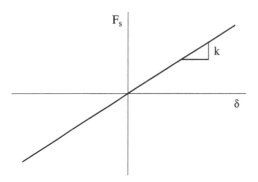

Figure 2.6 The linear spring constitutive relationship.

So long as $x_1 = x_2$, the spring is undeflected and simply moves as a rigid body. There will be no force in the spring unless $x_1 \neq x_2$, in which case the force will be a function of the change in length (i.e. displacement) of the spring, δ. The displacement must be consistent with the assumed direction of the force. Since we have chosen F_s such that it puts the spring in tension, then the spring must increase in length and x_2 must be bigger than x_1. The positive spring displacement is then $\delta = x_2 - x_1$.

The constitutive relationship for the linear spring element is plotted in Figure 2.6. The force in the spring can be expressed as

$$F_s = k\delta \tag{2.2}$$

where $\delta = y$ in this case since one end of the spring is fixed to the ground[3].

Substituting this into Equation 2.1 gives the equation of motion

$$m\ddot{y} + ky - mg = 0 \tag{2.3}$$

We now define the equilibrium value of y to be y_0 where the equilibrium condition is derived from Equation 2.3 by setting $\dot{y} = \ddot{y} = 0$, yielding

$$ky_0 - mg = 0 \tag{2.4}$$

This indicates that the spring has a "preload" in it that is exactly equal to the weight of the system. This is a universal fact. All systems that are in a stable equilibrium state have external forces on them that are exactly counteracted by internal forces. If this were not the case, there would be unbalanced forces causing accelerations and the system would not be in equilibrium. As we go on, it will become routine to derive equations governing small motions about the equilibrium state without including gravitational forces or preloads since we know they cancel each other out.

---

3 Rotate the spring in Figure 2.5 90 degrees CW and let the left-hand end of the spring be the end attached to ground. Then x_1 is zero and x_2 is equal to y. When modeling, it is advisable to make quick sketches of the various force producing elements and show how the motions of their ends are affected by the DOF definitions used in the model. Then the spring deflections or relative velocities across dampers can be quickly worked out.

We now let x be a small motion away from the equilibrium state so that we can write $y = y_0 + x$. Then, $\dot{y} = \dot{y}_0 + \dot{x} = \dot{x}$ since y_0 is a constant and $\ddot{y} = \ddot{x}$. Substituting these into Equation 2.3 yields

$$m\ddot{x} + kx + ky_0 - mg = 0 \tag{2.5}$$

where $ky_0 - mg = 0$ from the equilibrium condition in Equation 2.4, thereby giving the equation of motion as

$$m\ddot{x} + kx = 0 \tag{2.6}$$

Using Lagrange's Equation to find the equation of motion is straightforward. The kinetic energy is

$$T = \frac{1}{2}m\dot{y}^2 \tag{2.7}$$

and, taking the datum for gravitational potential energy to be at the bottom of the undeflected spring, the total potential energy is

$$U = \frac{1}{2}ky^2 - mgy \tag{2.8}$$

The terms needed for Lagrange's Equation are

$$q = y$$
$$\dot{q} = \dot{y}$$
$$\frac{\partial T}{\partial \dot{y}} = m\dot{y}$$
$$\frac{d}{dt}\left(\frac{\partial T}{\partial \dot{y}}\right) = m\ddot{y}$$
$$\frac{\partial T}{\partial y} = 0$$
$$\frac{\partial U}{\partial y} = ky - mg$$
$$Q_y = 0 \tag{2.9}$$

and substituting these expressions into Lagrange's Equation gives

$$m\ddot{y} + ky - mg = 0 \tag{2.10}$$

which is identical to Equation 2.3.

2.1.2 Equations of Motion Ignoring Preloads

We continue the modeling discussion of the previous section by looking at Figure 2.7, which is a repeat of Figure 2.1 except with gravity not being shown. x denotes a small displacement away from the equilibrium position. Figure 2.8 shows the free body diagram for the system. Note that this is not a complete free body diagram in the sense that both the gravitational load and the preload in the spring are not shown. Instead we treat the system as if the total spring deflection is equal to the displacement away from equilibrium. The resulting spring force shown on the FBD is then simply kx whereas we know that this is only the amount

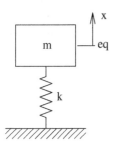

Figure 2.7 A mass on a spring.

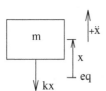

Figure 2.8 A mass on a spring – FBD.

that the spring force changes as the mass moves away from equilibrium and not the total spring force. The equation of motion, using Newton's Laws, is then

$$+\uparrow \sum F : -kx = m\ddot{x} \ \text{ or } \ m\ddot{x} + kx = 0 \tag{2.11}$$

which is exactly that derived in the previous section. To use Lagrange's Equation we would simply write

$$T = \frac{1}{2}m\dot{x}^2 \tag{2.12}$$

and

$$U = \frac{1}{2}kx^2 \tag{2.13}$$

which would quickly produce the same equation of motion.

The question of what forces can be left out of the analysis is best answered by saying that *if the effect of the force changes direction when the system passes through equilibrium then it must be kept*. The gravitational force acting on the mass in the system being considered here is always pulling the mass down regardless of whether it is above or below the equilibrium position. Since it doesn't change direction, it can be left out of the analysis. We can make a similar statement about the preload – it always opposes the gravitational load so it never changes direction and can be left out. If, on the other hand, we are deriving the equation of motion for the simple pendulum shown in Figure 2.9, then we see that the gravitational force has a different effect on either side of the equilibrium position. If the pendulum swings to the right (positive θ) then gravity pulls it to the left and, if the pendulum swings to the left (negative θ), gravity pulls it to the right. In this case, therefore, the gravitational force must remain in the equation of motion for the pendulum. Forces that change direction when the system passes through equilibrium are generally called *restoring forces* and are included in the equation of motion[4].

---

4 Springs are the most common elements that we think of as providing a restoring force since they are always trying to bring the system back to its stable equilibrium state. However, the name "restoring force"

Figure 2.9 A simple pendulum.

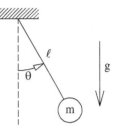

2.1.3 Finding Spring Deflections due to Body Rotations

Very often vibration models have bodies whose rotation causes spring deflections. In these cases, the springs are placed at a distance from the axis of rotation and the forces induced in them through the body rotation cause *restoring moments* that attempt to return the body to its equilibrium state. Figure 2.10 shows the geometry involved in the spring deflection for a large angle of rotation, θ.

On the left of the figure is shown a rigid rod pinned to the ground at A and attached to the ground, a distance d away from the pivot point, by a spring that has a stiffness, k, and a free, or undeflected, length of ℓ_o. On the right, the same system is shown after a large rotation about point A. As you can see, the spring is now longer by virtue of having undergone a deflection, δ. The spring has also rotated slightly away from its original vertical orientation by a clockwise angle, β.

We can write two nonlinear algebraic equations for the horizontal and vertical distances from A to C. Since A and C are fixed in space, the distance between them is constant. The horizontal distance, from the left-hand image, is d and, from the right-hand image, is $d \cos \theta + (\ell_o + \delta) \sin \beta$. Equating these distances gives

$$d = d \cos \theta + (\ell_o + \delta) \sin \beta \tag{2.14}$$

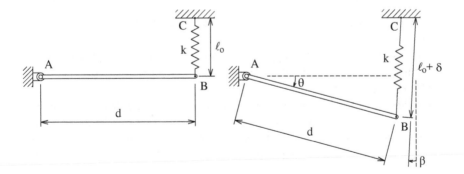

Figure 2.10 Spring deflection due to a large angle of rotation.

---

is used for all forces that change direction when passing through equilibrium even though the force may, in some cases, try to move the system away from equilibrium. Consider, for instance, the gravitational effect on an inverted pendulum held in place by elastic elements. Gravity introduces a negative stiffness in this case but it still must be retained in the equation of motion.

Similarly, the vertical distance is ℓ_o from the left-hand image and $-d\sin\theta + (\ell_o + \delta)\cos\beta$ from the right-hand image. This gives

$$\ell_o = -d\sin\theta + (\ell_o + \delta)\cos\beta \tag{2.15}$$

For a system with given dimensions and a specified angle of rotation, θ, Equations 2.14 and 2.15 become a set of two simultaneous algebraic equations with two unknowns, β and δ. They can be solved in the following way. First, rearrange them as

$$(\ell_o + \delta)\sin\beta = d - d\cos\theta$$
$$(\ell_o + \delta)\cos\beta = \ell_o + d\sin\theta \tag{2.16}$$

Now divide the first of Equation 2.16 by the second to find

$$\tan\beta = \frac{d(1 - \cos\theta)}{\ell_o + d\sin\theta} \tag{2.17}$$

This is not pretty but, given values for d, ℓ_o and θ, we could use the inverse tangent function to find a numerical value for the angle, β.

Then we can proceed to find δ as follows. Multiply Equation 2.14 by $\sin\beta$ everywhere to get

$$d\sin\beta = d\cos\theta\sin\beta + (\ell_o + \delta)\sin^2\beta \tag{2.18}$$

Then multiply Equation 2.15 by $\cos\beta$, giving

$$\ell_o\cos\beta = -d\sin\theta\cos\beta + (\ell_o + \delta)\cos^2\beta \tag{2.19}$$

These last two equations can be added together to give

$$d\sin\beta + \ell_o\cos\beta = d(\cos\theta\sin\beta - \sin\theta\cos\beta) + (\ell_o + \delta)(\sin^2\beta + \cos^2\beta) \tag{2.20}$$

after which we can make use of two trigonometric identities and rearrange to get

$$\delta = d\sin\beta + d\sin(\theta - \beta) - \ell_o(1 - \cos\beta) \tag{2.21}$$

which is, again, not pretty but permits numerical values of δ to be calculated given that we know the dimensions and have already solved for β.

These solutions would be perfectly adequate in a nonlinear numerical simulation of the system but that isn't our goal. For a vibrational analysis, we limit ourselves to small motions about the stable equilibrium state, so let's do that here.

Assume that the left-hand image is the stable equilibrium state and that θ is small. For small values of θ, Equation 2.17 indicates that $\tan\beta = 0$ or, in fact, $\beta = 0$. In other words, the spring has no motion perpendicular to its line of action for small rotations. Substituting $\beta = 0$ into Equation 2.21 gives the spring displacement, for small θ,

$$\delta = d\theta \tag{2.22}$$

All of this work has led us to a very simple result for the small rotational degrees of freedom used in vibrations. *The point where the spring is connected to the body experiences a total motion that is equal in magnitude to the arc length at that point resulting from the small rotation about the axis of rotation and is in a direction perpendicular to the radius vector from the axis of rotation to the point where the spring is connected to the body.*

Figure 2.11 A body with a rotational DOF: How to deal with a spring, a damper, and gravitational effects.

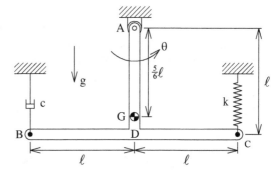

Consider another system, that shown in Figure 2.11. The body shown in the figure is constructed of three identical, thin, uniform, rods (AD, BD, and CD) that are welded together at D. Each rod has mass, m, and length, ℓ. The body is suspended from a frictionless pivot at A and acts as a pendulum since the center of mass is located a distance $5\ell/6$ below point A[5]. There is a vertical spring (stiffness $= k$) at C and a vertical damper (damping coefficient $= c$) at B that resist motion of the body.

Consider first the deflection of the spring at C. Let the degree of freedom be a small counterclockwise rotation, θ, about point A. Using our arc length analysis, we note that point C will move up and to the right. It will move a distance $\ell\theta$ to the right because of the vertical link AD and a distance $\ell\theta$ upward because of the horizontal link DC. It is important to note that the spring responds only to deflections aligned with it and, since this is a vertical spring, the horizontal motion has no effect other than to make the spring rotate about its point of connection to the ground. It is free to make this rotation without resistance of any kind. As a result, only the bottom of the spring moving upward a distance $\ell\theta$ causes a resisting force. Since the top of the spring is connected to the ground, it has no motion so the spring gets shorter and is therefore in compression with a deflection equal to $\ell\theta$. The resulting force in the spring is $F_s = k\ell\theta$. This must show up on the FBD of the body. It will be perpendicular to DC and will be pushing downward since it is a compressive force.

To derive the equation of motion using Lagrange's Equation, we add a potential energy term

$$U = \frac{1}{2}k(\ell\theta)^2$$

where you can see that all nonlinearities have been left out and we are going directly for the linear equation of motion. This is a much more convenient way of deriving equations of motion than our previous method, introduced in Chapter 1, where we included all the nonlinear terms, found equilibrium, and then linearized for small motions about equilibrium. Here, we know that the system is in equilibrium and define small motions about that state, accounting for the changes in forces away from their equilibrium values only. While being convenient, there are hidden dangers in taking this shortcut. If you are ever uncomfortable about your derived equation of motion, I recommend going back to the original method and taking a few minutes to confirm that you get the same result both ways.

---

5 Being able to find the location of the center of mass of a composite body is a useful skill and one worth practicing.

The damper at B is treated in a manner similar to that we used on the spring. We first consider the motion of point B – it moves to the right and downward as the counterclockwise small rotation, θ, occurs. Since the damper is vertical, it will only respond to the vertical component of the total motion. That component is $\ell\theta$ downward. The damper is fixed to the ground at its top and its bottom is moving downward so it is getting longer and is therefore in tension, implying that it will pull upward on the body during the motion. The vertical deflection of the damper is $\ell\theta$ but, since the damper's constitutive relationship is

$$F_d = cv_{rel}$$

where c is the damping coefficient and v_{rel} is the relative velocity between the ends of the damper, we need to differentiate the deflection to get the velocity of the end of the damper that is connected to the body. This gives $v_{rel} = \ell\dot{\theta}$ and therefore

$$F_d = c\ell\dot{\theta}$$

pulling vertically upward on the body at point B.

To derive the equation of motion using Lagrange's Equation, we add a Rayleigh's dissipation function term (see Subsection 1.1.3.4)

$$\mathfrak{R} = \frac{1}{2}c(\ell\dot{\theta})^2$$

where we are, once again, considering only the small deviations from equilibrium.

Putting gravitational effects into the linearized equations of motion directly takes a little more thought and effort. Using Newton's Laws is fairly straightforward but we still need to use trigonometric functions before getting the linearized form. We first show the gravitational force, $F_g = 3mg$, acting vertically downward through the center of mass and then rotate the body through a large enough angle so that we can see a horizontal moment arm developing between the gravitational force and the pivot point at A (see Figure 2.12). The moment arm is $\frac{5}{6}\ell\sin\theta$, causing the moment about A from the gravitational force to be $3mg\left(\frac{5}{6}\ell\sin\theta\right)$ in the clockwise direction. This can be linearized to $3mg\left(\frac{5}{6}\ell\right)\theta$.

Using Lagrange's Equations to include gravitational effects due to rotational motion always requires a nonlinear analysis followed by linearization because there is a derivative of the potential energy with respect to the degree of freedom that enters the equation of

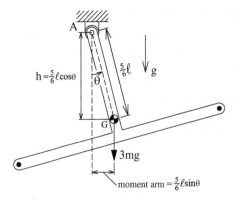

Figure 2.12 Gravitational effects.

motion. For this problem, for example, we will need to include $\partial U/\partial \theta$ in the equation of motion, where U is the potential energy and includes both gravitational and elastic components. The gravitational component, in this case and for a datum at A, is $U_g = -3mgh$ where, from Figure 2.12, $h = \frac{5}{6}\ell \cos\theta$, giving $U_g = -3mg\left(\frac{5}{6}\ell\right)\cos\theta$.

If we linearize the potential energy before taking the derivative, we substitute the usual approximation, $\cos\theta = 1$, and arrive at $U_g = -3mg\left(\frac{5}{6}\ell\right)$ which is clearly a constant and therefore has a derivative equal to zero. To get the correct formulation of the equation of motion, *it is imperative to take the derivative of the gravitational potential energy while it is still nonlinear and then linearize the result*[6]. In this case,

$$\frac{\partial U_g}{\partial \theta} = -3mg\left(\frac{5}{6}\ell\right)\frac{\partial(\cos\theta)}{\partial\theta} = 3mg\left(\frac{5}{6}\ell\right)\sin\theta$$

as before. This can be linearized to exactly what we got with Newton's Laws: $3mg\left(\frac{5}{6}\ell\right)\theta$.

We finish this section by using Lagrange's Equation to get the equation of motion.

The kinetic energy of a rigid body rotating about a fixed point can be expressed as one half of the moment of inertia about that point multiplied by the square of the angular velocity of the body. In this case, the moment of inertia[7] about point A is $I_A = 3m\ell^2$. The kinetic energy is then

$$T = \frac{1}{2}(3m\ell^2)\dot\theta^2 \tag{2.23}$$

The potential energy is the sum of the elastic and gravitational terms that we have already derived.

$$U = \frac{1}{2}k(\ell\theta)^2 - 3mg\left(\frac{5}{6}\ell\right)\cos\theta \tag{2.24}$$

The damping term requires Rayleigh's Dissipation function, which we showed to be

$$\mathfrak{R} = \frac{1}{2}c(\ell\dot\theta)^2 \tag{2.25}$$

All that remains is to substitute into Lagrange's Equation as presented in Chapter 1, Equation 1.43 and repeated here.

$$\frac{d}{dt}\left(\frac{\partial T}{\partial \dot{q}_r}\right) - \frac{\partial T}{\partial q_r} + \frac{\partial U}{\partial q_r} + \frac{\partial \mathfrak{R}}{\partial \dot{q}_r} = Q_{q_r} \tag{2.26}$$

---

6 In Chapter 1, we used a Maclaurin's Series expansion of the cosine function when we discussed linearization. The expansion is

$$\cos\theta = 1 - \frac{\theta^2}{2!} + \frac{\theta^4}{4!} - \frac{\theta^6}{6!} + \cdots$$

Linearization of the cosine involves a truncation of this series after the first term, leading to our familiar approximation for small angles, $\cos\theta \approx 1$. As we are going to differentiate the cosine function here, we could simply keep the first two terms of the series

$$\cos\theta \approx 1 - \frac{\theta^2}{2!}$$

and the derivative with respect to theta would give the linear result we seek.

7 To repeat an earlier footnote about being able to find the location of the center of mass of a composite body, being able to find the moment of inertia of a composite body is also a useful skill and one worth practicing. Refer to Appendix B for some help with this.

We find

$$q_r = \theta$$

$$\dot{q}_r = \dot{\theta}$$

$$\frac{\partial T}{\partial \dot{\theta}} = 3m\ell^2 \dot{\theta}$$

$$\frac{d}{dt}\left(\frac{\partial T}{\partial \dot{\theta}}\right) = 3m\ell^2 \ddot{\theta}$$

$$\frac{\partial T}{\partial \theta} = 0$$

$$\frac{\partial U}{\partial \theta} = k\ell^2\theta + 3mg\left(\frac{5}{6}\ell\right)\sin\theta$$

$$\frac{\partial \mathfrak{R}}{\partial \dot{\theta}} = c\ell^2\dot{\theta}$$

$$Q_\theta = 0 \tag{2.27}$$

Putting this all together with some rearranging and a linearization of $\sin\theta$ to be simply θ leads to the equation of motion

$$3m\ell^2\ddot{\theta} + c\ell^2\dot{\theta} + \left(k\ell^2 + \frac{5}{2}mg\ell\right)\theta = 0 \tag{2.28}$$

Deriving this equation using Newton's Laws is left as an exercise below.

Exercises

2.1 Use Newton's method to derive the equation of motion for the system in Figure 2.11 and show that it is the same as that given in Equation 2.28.

2.2 The two particle masses (m and $2m$) are attached to each other by the rigid massless link shown. The link is pinned to the ground with a frictionless pin at point O and a horizontal damper c, below the pin, connects the link to ground.
 (a) Use Newton's laws to write the equation of motion for the system using a small angle of rotation, θ, around the equilibrium position shown as the degree of freedom.
 (b) Use Lagrange's Equation to derive the equation of motion and confirm that the two derivations give the same result.

2.3 The pulleys in the figure are massless and frictionless and the cord is massless and inextensible. Let y denote the motion of the mass downward from the position where the spring is undeflected.
 (a) Use Newton's Laws to write the equation of motion for the system.[8]

---

8 The *inextensible* cord gives a constraint on the relative motion between bodies because it cannot change in length. To determine how the vertical motion of the top of the damper is related to the vertical motion of the mass, let the total length of the cord be L, a constant, where L is equal to the sum of the lengths of the

Figure E2.2

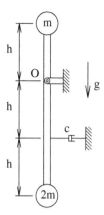

(b) Find the equilibrium condition and solve for y_0, the equilibrium value of y.

(c) Let there be a small motion, x, away from equilibrium so that $y = y_0 + x$ and write the resulting equation of motion for small motions about equilibrium.

Figure E2.3

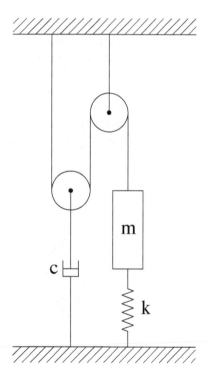

cord on the left, the cord in the center, the cord on the right, and the cord wrapped around the two pulleys (it helps to give these individual lengths names like ℓ_1, ℓ_2, and so on). Let the mass move down a distance, y, and the top of the damper move up a distance, z. Now rewrite the sum of all the lengths where the length on the left and in the center have both decreased by z and the length on the right has increased by y. The total is still equal to L so you can equate the two expressions for L to see how the two motions are related to each other.

2.4 The mass m is supported by an inextensible, massless, cable that passes over two, frictionless, massless pulleys before being connected to ground. The lower pulley is attached to a rigid, massless, lever of length $2d$. The lever pivots about point O without friction and is attached to ground via a spring (stiffness $= k$) at its midpoint and a damper (damping coefficient $= c$) at its endpoint. Derive the equation of motion twice and show that both methods give the same result.

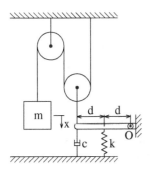

Figure E2.4

2.5 The circular disk shown in the figure rotates in the horizontal plane[9] with constant angular speed, ω. A slot has been milled into the disk in a radial direction as shown. The small mass, m, is able to move along the slot without friction and is attached to the center of the disk by a spring of stiffness, k. The spring has an undeflected length of ℓ_0 (i.e. there is no force in the spring when it has a length of ℓ_0).

(a) Consider the case where the disk is not rotating (i.e. $\omega = 0$). Write the equation of motion for small motions of the mass as it moves away from its equilibrium position at the undeflected length of the spring.

Figure E2.5

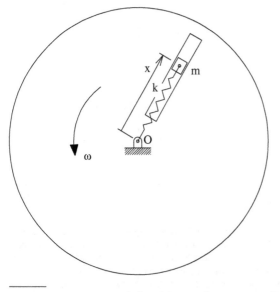

---

9 Rotating in the *horizontal plane* is just another way to say "ignore gravitational effects".

(b) Consider now the case where the disk has a constant angular speed and the mass is resting at a constant distance from the center of the disk. What is the distance from the center of the disk to the mass in this equilibrium state?

(c) Consider finally the case where the mass is slightly disturbed from its equilibrium position in part (b). The angular speed of the disk is unchanged. Write the equation of motion for the mass and determine the effective stiffness. For large enough angular speeds, this stiffness can be negative. What does that say about the behavior of the system for large angular speeds?

2.6 Three identical uniform slender rods are attached to each other and to the ground with frictionless pinned joints as shown in the figure. Each rod has mass m and length ℓ. The system behaves as a pendulum when it is given the small angular displacement θ and released.

(a) Write an expression for the kinetic energy of the system as a function of $\dot{\theta}$.

(b) Write an expression for the potential energy of the system as a function of θ.

(c) Use Lagrange's Equation to derive the equation of motion for the system.

Figure E2.6

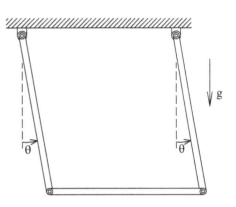

3

Single Degree of Freedom Systems – Free Vibrations

The equations of motion derived in Chapter 2 were the *free vibration* equations of motion for systems. Free vibration equations of motion always have the form

$$m\ddot{x} + c\dot{x} + kx = 0 \tag{3.1}$$

where m represents the *effective mass*, c represents the *effective damping,* and k represents the *effective stiffness* of the system. Remember that we are considering systems like the independent front suspension described earlier (see Figure 2.2) for the most part and the values of m, c, and k are not as simple as those shown in the standard mass/spring/damper systems that we normally use for illustrating vibratory systems. At the end of Chapter 2, we derived an equation of motion for a system that appeared as

$$3m\ell^2\ddot{\theta} + c\ell^2\dot{\theta} + \left(k\ell^2 + \frac{5}{2}mg\ell \right)\theta = 0 \tag{3.2}$$

In this equation, the effective mass is $3m\ell^2$, the effective damping is $c\ell^2$, and the effective stiffness is $\left(k\ell^2 + \frac{5}{2}mg\ell \right)$. The degree of freedom is θ rather than x but Equations 3.1 and 3.2 will have solutions that are identical in form.

The left- and right-hand sides of the equation of motion have different physical meanings. The left-hand side contains all of the system properties – mass, stiffness, and damping. The right-hand side indicates what we do to the system – apply forces. At this point the right-hand side is zero, indicating that we have not applied a force to the system. We call the motions of systems like this *free vibrations* since the systems are free to do what they want without any external interference. In this chapter, we consider free vibrations of systems, realizing that motion can only ensue if there is a set of initial conditions that starts the motion.

3.1 Undamped Free Vibrations

Setting the effective damping coefficient, c, to zero in Equation 3.1 gives the equation of motion for undamped free vibration. It is

$$m\ddot{x} + kx = 0 \tag{3.3}$$

Introduction to Mechanical Vibrations, First Edition. Ronald J. Anderson.
© 2020 John Wiley & Sons Ltd. Published 2020 by John Wiley & Sons Ltd.
Companion website: www.wiley.com/go/anderson/introduction-to-vibrations

This is a second-order, linear, ordinary, differential equation and we should be able to recall enough about our differential equations course to solve it. However, let's put aside any previous knowledge of solving differential equations and just look at the equation and see what we expect from it.

We know that we have a physical system and that, if we move it away from a stable equilibrium state and then release it, the restoring forces will try to move it back. The motion will be characterized by a function, $x(t)$, which will be a measure of how far the mass is from its equilibrium position as time goes on and $x(t)$ must satisfy Equation 3.3. The requirement that it satisfy Equation 3.3 severely restricts the form of the function, $x(t)$, because it must be some function whose second derivative looks just like $x(t)$. That is, we must be able to multiply $x(t)$ by a constant, k, and then add its second derivative, $\ddot{x}(t)$, multiplied by another constant, m, and get a result of zero for all values of time. One requirement for this to happen is that the function repeat itself on its second derivative. Furthermore, the second derivative must have the opposite sign to the original function since both m and k are positive and there needs to be a sign change in order to make the two terms in Equation 3.3 sum to zero.

There are only three functions that can satisfy these requirements. They are, for a constant, p, to be determined:

- The trigonometric function $\cos pt$. The derivatives are

$$\frac{d}{dt}\cos pt = -p\sin pt$$

and

$$\frac{d^2}{dt^2}\cos pt = -p^2\cos pt$$

- The trigonometric function $\sin pt$. The derivatives are

$$\frac{d}{dt}\sin pt = p\cos pt$$

and

$$\frac{d^2}{dt^2}\sin pt = -p^2\sin pt$$

- The exponential function e^{ipt} where $i = \sqrt{-1}$. The derivatives are

$$\frac{d}{dt}e^{ipt} = ipe^{ipt}$$

and

$$\frac{d^2}{dt^2}e^{ipt} = i^2p^2e^{ipt} = -p^2e^{ipt}$$

Given that we are working with a linear differential equation; it is possible to superimpose solutions so a linear combination of the three possible solutions is also a solution. We can let $x(t)$ be

$$x(t) = A\cos pt + B\sin pt + Ce^{ipt} \tag{3.4}$$

where A, B, and C are constants. We also need to remember Euler's Identity

$$e^{ipt} = \cos pt + i\sin pt \tag{3.5}$$

which means that Equation 3.4 can be written as

$$x(t) = (A + C)\cos pt + (B + iC)\sin pt \tag{3.6}$$

In other words, the two trigonometric functions taken together carry the same information as the exponential function by itself. We will see as we move forward that it is much easier to use the complex exponential notation in most cases, especially where there is damping.

For the case of undamped, free vibrations we can choose any one of the three functions to get the most pressing result. That is, we want to determine the constant p. To do this, let

$$x(t) = A \cos pt \tag{3.7}$$

from which

$$\ddot{x}(t) = -p^2 A \cos pt \tag{3.8}$$

Substituting into Equation 3.3 yields

$$-mp^2 A \cos pt + kA \cos pt = 0 \tag{3.9}$$

or

$$(-mp^2 + k)A \cos pt = 0 \tag{3.10}$$

The goal is to satisfy Equation 3.10 for all time, t. We can state outright that $\cos pt$ is not zero at all times so it can't be used to satisfy the equation. $A = 0$ is a solution but A is the amplitude of the motion and setting it to zero means the system stays in equilibrium forever and that isn't the solution we want. The only option that permits motion while satisfying Equation 3.10 is to let

$$(-mp^2 + k) = 0 \tag{3.11}$$

and, since m and k are known constants, this requires that

$$p = \pm\sqrt{\frac{k}{m}} \tag{3.12}$$

Equation 3.12 is one of the most important results in the field of vibrations. The constant, p, is given the name *undamped natural frequency* and the symbol ω_n. The functional expression for $x(t)$ is then

$$x(t) = A \cos \omega_n t \tag{3.13}$$

and we can see that the *natural motion* of the undamped mass/spring system, once disturbed from equilibrium, is a harmonic function of time that repeats itself whenever $\omega_n t$ is a multiple of 2π. The amplitude is constant and the motion is oscillatory with a period T derived from $\omega_n T = 2\pi$, yielding $T = 2\pi/\omega_n$. This motion is called Simple Harmonic Motion or SHM and is depicted in Figure 3.1.

3.2 Response to Initial Conditions

The natural response of the undamped system just described lets us know what the system wants to do but not exactly what it will do. In fact, we were able to find the natural

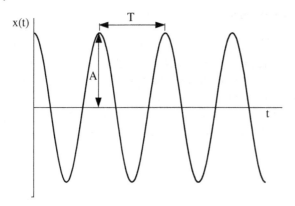

Figure 3.1 Simple harmonic motion.

frequency and we said there was a non-zero amplitude, A, but we can't exactly say what value A has unless we know how the motion was started. The motion is started by a set of *initial conditions*. The initial conditions are an initial displacement, $x(0) = x_0$, and an initial velocity, $\dot{x}(0) = v_0$. You can visualize these conditions as the act of pulling the mass away from equilibrium in such a way that, just as it is passing through the initial displacement, x_0, it has a velocity, v_0, and you release it with these starting conditions.

The general form of the response will be a superposition of $\cos \omega_n t$ and $\sin \omega_n t$ that can be written as

$$x(t) = A \cos \omega_n t + B \sin \omega_n t \tag{3.14}$$

where A and B are constants to be determined from the initial conditions.

We differentiate Equation 3.14 to get an expression for the velocity

$$\dot{x}(t) = -\omega_n A \sin \omega_n t + \omega_n B \cos \omega_n t \tag{3.15}$$

and then set the displacement and velocity at time $t = 0$ to x_0 and v_0 respectively to get

$$x_0 = A \cos 0 + B \sin 0 \tag{3.16}$$

and

$$v_0 = -\omega_n A \sin 0 + \omega_n B \cos 0 \tag{3.17}$$

Noting that $\cos 0 = 1$ and $\sin 0 = 0$. the result is that the constants are $A = x_0$ and $B = v_0/\omega_n$ so that the displacement is

$$x(t) = x_0 \cos \omega_n t + \frac{v_0}{\omega_n} \sin \omega_n t \tag{3.18}$$

Since there are two terms in Equation 3.18, it isn't clear what the actual amplitude of the motion is. In fact, the initial velocity has introduced a *phase shift* in the result. You can see the effect by setting $v_0 = 0$ in Equation 3.18 with the result that

$$x(t) = x_0 \cos \omega_n t \tag{3.19}$$

which is exactly the function we assumed in Equation 3.7 and plotted in Figure 3.1.

With a non-zero initial velocity, we can write

$$x(t) = X \cos(\omega_n t + \phi) \tag{3.20}$$

where we find the amplitude, X, and the phase angle, ϕ, by looking at the initial conditions. That is, setting $t = 0$, we write the initial displacement as

$$X \cos \phi = x_0 \tag{3.21}$$

Then we differentiate to find an expression for the velocity

$$\dot{x}(t) = -\omega_n X \sin(\omega_n t + \phi) \tag{3.22}$$

and set this equal to the initial velocity when $t = 0$.

$$-\omega_n X \sin \phi = v_0 \tag{3.23}$$

from which we find

$$X \sin \phi = -\frac{v_0}{\omega_n} \tag{3.24}$$

We can square Equations 3.21 and 3.24 and add them together to get

$$X^2 \cos^2 \phi + X^2 \sin^2 \phi = X^2 = x_0^2 + \frac{v_0^2}{\omega_n^2} \tag{3.25}$$

from which the amplitude is seen to be

$$X = \sqrt{x_0^2 + \frac{v_0^2}{\omega_n^2}} \tag{3.26}$$

Equation 3.24 can be divided by Equation 3.21, giving

$$\frac{X \sin \phi}{X \cos \phi} = \tan \phi = -\frac{v_0}{\omega_n x_0} \tag{3.27}$$

from which the phase angle is found to be

$$\phi = -\tan^{-1}\left(\frac{v_0}{\omega_n x_0}\right) \tag{3.28}$$

Figure 3.2 shows how the phase shift affects the response of the system. The simple harmonic motion is still present but the signal starts at time $t = 0$ with the initial displacement, x_0, and the initial velocity, v_0, which is the slope of the curve at $t = 0$. The amplitude is X

Figure 3.2 Simple harmonic motion with a phase shift.

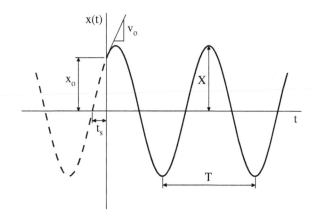

from Equation 3.26 and the phase angle can be found from the ratio of t_s, a measure of where the displacement would have been zero, to the period T. That is

$$\phi = -2\pi \frac{t_s}{T} \tag{3.29}$$

3.3 Damped Free Vibrations

We now consider the response of a system that has mass, stiffness, and damping as shown in Figure 3.3. The free body diagram for the system is shown in Figure 3.4 and the force balance equation is

$$+\uparrow \sum F : -kx - c\dot{x} = m\ddot{x} \text{ or } m\ddot{x} + c\dot{x} + kx = 0 \tag{3.30}$$

The solution, $x(t)$, must now be some function that repeats itself on every derivative. That is, we want to add together $x(t)$ and its first two derivatives and have them sum to zero. This can only happen if they all are the same function of time. The only function that has this property is the exponential function. We therefore let

$$x(t) = Xe^{\lambda t} \tag{3.31}$$

where X and λ are undetermined constants.

The derivatives are

$$\dot{x}(t) = \lambda Xe^{\lambda t} \tag{3.32}$$

and

$$\ddot{x}(t) = \lambda^2 Xe^{\lambda t} \tag{3.33}$$

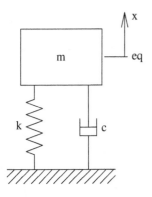

Figure 3.3 Mass/spring/damper system.

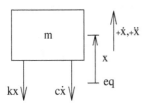

Figure 3.4 Mass/spring/damper FBD.

Substituting Equations 3.31 through 3.33 into Equation 3.30 gives

$$(m\lambda^2 + c\lambda + k)Xe^{\lambda t} = 0 \tag{3.34}$$

as the requirement for a solution. Considering this equation, we can rule out $e^{\lambda t} = 0$ since the exponential function is never zero and, also, $X = 0$ since that would mean no motion whatsoever. As a result, the only way to have any motion is to have

$$m\lambda^2 + c\lambda + k = 0 \tag{3.35}$$

Equation 3.35 is a quadratic polynomial with two roots and is called the *characteristic equation* for the system. The roots[1] are

$$\lambda = \frac{-c \pm \sqrt{c^2 - 4mk}}{2m} \tag{3.36}$$

The fact that there are two roots, call them λ_1 and λ_2, means that the original assumption of an exponential solution to the linear differential equation needs to be expanded to superimpose the two solutions. We therefore write the complete solution as

$$x(t) = X_1 e^{\lambda_1 t} + X_2 e^{\lambda_2 t} \tag{3.37}$$

where X_1 and X_2 need to be found from initial conditions but we will leave that until later.

The square root in Equation 3.36 leads to interesting system behavior. There is the possibility of the term inside the square root (i.e. $c^2 - 4mk$) being positive, zero, or negative.

If it is positive, then there will be two negative real roots[2]. In this case, $x(t)$ will simply consist of two decaying exponentials without any oscillations. This is called the *overdamped* case because the damping is too large to allow oscillations. In fact, the two exponentials decay at different rates with the least negative root decaying more slowly than the other. The result is that the response is dominated by the root nearest the origin.

If it is zero, there are two identical negative real roots

$$\lambda_1 = \lambda_2 = -\frac{c}{2m} \tag{3.38}$$

and the solution has to be rewritten to make the two exponential solutions linearly independent. The response is now

$$x(t) = (X_1 + X_2 t)e^{-(c/2m)t} \tag{3.39}$$

---

1 Use the equation for the roots of a quadratic equation. Given

$$ax^2 + bx + c = 0$$

then the roots are

$$x = \frac{-b \pm \sqrt{b^2 - 4ac}}{2a}$$

2 The roots are

$$-\frac{c}{2m} \pm \sqrt{\left(\frac{c}{2m}\right)^2 - \frac{k}{m}}$$

The two roots are to the left and right of $-c/2m$ and are equidistant from it. The distance from $-c/2m$ to either root is the square root term and it can never be larger than $c/2m$ so the root to the right can never reach zero or be positive.

and has the fastest decay time of any of the solutions. This is called *critical damping* and we define a critical damping coefficient, c_{cr}, as the value of c that makes $c^2 - 4mk = 0$.

That is

$$c_{cr} = \sqrt{4mk} \qquad (3.40)$$

The third possibility is that $c^2 - 4mk < 0$. In this case, the solution involves the square root of a negative number so that the two roots are complex conjugates that can be written as

$$\lambda_1 = -\frac{c}{2m} - i\sqrt{\frac{k}{m} - \left(\frac{c}{2m}\right)^2} \qquad (3.41)$$

and

$$\lambda_1 = -\frac{c}{2m} + i\sqrt{\frac{k}{m} - \left(\frac{c}{2m}\right)^2} \qquad (3.42)$$

This is the *underdamped* case where damping is small enough that oscillations occur but the amplitudes of the oscillations are decaying exponentially.

While we can describe the type of motion that exists for the three cases of overdamped, critically damped, and underdamped systems using the equations with the three physical parameters m, c, and k, the mathematical form of the solutions is much enhanced by first changing the equation of motion to the *standard form* for a second-order system where only two parameters are employed.

3.3.1 Standard Form for Second-Order Systems

We begin with the equation of motion for damped free vibrations with one degree of freedom

$$m\ddot{x} + c\dot{x} + kx = 0 \qquad (3.43)$$

and divide throughout by the mass to get

$$\ddot{x} + \frac{c}{m}\dot{x} + \frac{k}{m}x = 0 \qquad (3.44)$$

We recognize k/m to be the undamped natural frequency ω_n squared (see Equation 3.12 and its following paragraph).

The term c/m needs a little work. We start by defining a new parameter called the *damping ratio*. This is the ratio of the actual damping in the system to the critical damping for the system. We represent this parameter with the Greek letter zeta (ζ).

$$\zeta = \frac{c}{c_{cr}} \qquad (3.45)$$

The three types of system response can now be classified using ζ.

- If $\zeta > 1$, then the actual damping is greater than critical damping and the system is overdamped.
- If $\zeta = 1$, then the actual damping is equal to the critical damping and the system is critically damped.

- If $\zeta < 1$, then the actual damping is smaller than critical damping and the system is underdamped.

We can write c/m as

$$\frac{c}{m} = \frac{c}{c_{cr}}\frac{c_{cr}}{m} = \zeta\frac{c_{cr}}{m} \tag{3.46}$$

and then substitute $c_{cr} = \sqrt{4mk}$ from Equation 3.40 to get

$$\frac{c}{m} = \zeta\frac{\sqrt{4mk}}{m} = 2\zeta\sqrt{\frac{k}{m}} = 2\zeta\omega_n \tag{3.47}$$

Then, Equation 3.44 can be rewritten as

$$\ddot{x} + 2\zeta\omega_n\dot{x} + \omega_n^2 x = 0 \tag{3.48}$$

where we see that the system is characterized by only two parameters – the undamped natural frequency and the damping ratio. We will see that these two parameters will be used to describe even the most complicated systems we address in the field of vibrations.

We assume again that the solutions are exponential in nature

$$x(t) = Xe^{\lambda t} \tag{3.49}$$

and substitute into Equation 3.48 to get

$$(\lambda^2 + 2\zeta\omega_n\lambda + \omega_n^2)Xe^{\lambda t} = 0 \tag{3.50}$$

from which the characteristic equation is

$$\lambda^2 + 2\zeta\omega_n\lambda + \omega_n^2 = 0 \tag{3.51}$$

with roots

$$\lambda = \frac{-2\zeta\omega_n \pm \sqrt{4\zeta^2\omega_n^2 - 4\omega_n^2}}{2} = -\zeta\omega_n \pm \omega_n\sqrt{\zeta^2 - 1} \tag{3.52}$$

We can now consider the values of the roots for four different cases.

3.3.2 Undamped

When $\zeta = 0$, the system is undamped and the roots from Equation 3.52 are

$$\lambda_1 = -i\omega_n \;;\; \lambda_2 = +i\omega_n$$

with the solution being

$$x(t) = X_1 e^{-i\omega_n t} + X_2 e^{+i\omega_n t} \tag{3.53}$$

which, using Euler's Identity, can be written as

$$x(t) = X_1(\cos\omega_n t - i\sin\omega_n t) + X_2(\cos\omega_n t + i\sin\omega_n t) \tag{3.54}$$

or

$$x(t) = (X_1 + X_2)\cos\omega_n t - i(X_1 - X_2)\sin\omega_n t \tag{3.55}$$

We differentiate to find an expression for the velocity so that we can apply initial conditions

$$\dot{x}(t) = -\omega_n (X_1 + X_2) \sin \omega_n t - i\omega_n (X_1 - X_2) \cos \omega_n t \qquad (3.56)$$

Then, for $x(0) = x_0$ and $\dot{x}(0) = v_0$, we find

$$x(0) = (X_1 + X_2) \cos 0 - i(X_1 - X_2) \sin 0 = x_0 \qquad (3.57)$$

and

$$\dot{x}(0) = -\omega_n (X_1 + X_2) \sin 0 - i\omega_n (X_1 - X_2) \cos 0 = v_0 \qquad (3.58)$$

Equations 3.57 and 3.58 respectively yield $(X_1 + X_2) = x_0$ and[3] $-i(X_1 - X_2) = v_0/\omega_n$. When substituted into Equation 3.55 we find the undamped response to initial conditions as

$$x(t) = x_0 \cos \omega_n t + \frac{v_0}{\omega_n} \sin \omega_n t \qquad (3.59)$$

which is identical to what we found previously in Equation 3.18.

3.3.3 Underdamped

When $0 < \zeta < 1$, the system is underdamped and the roots from Equation 3.52 are

$$\lambda_1 = -\zeta\omega_n - i\omega_n \sqrt{1 - \zeta^2} \ ; \ \lambda_2 = -\zeta\omega_n + i\omega_n \sqrt{1 - \zeta^2}$$

with the superimposed exponential solution

$$x(t) = X_1 e^{(-\zeta\omega_n - i\omega_n \sqrt{1-\zeta^2})t} + X_2 e^{(-\zeta\omega_n + i\omega_n \sqrt{1-\zeta^2})t} \qquad (3.60)$$

At this point we define a new frequency – the *damped natural frequency*, ω_d, as

$$\omega_d = \omega_n \sqrt{1 - \zeta^2} \qquad (3.61)$$

and rewrite $x(t)$ as

$$x(t) = e^{-\zeta\omega_n t} [X_1 e^{-i\omega_d t} + X_2 e^{+i\omega_d t}] \qquad (3.62)$$

where we have factored out the common term

$$e^{-\zeta\omega_n t}$$

as well as inserting the damped natural frequency.

---

3 It may seem odd that we are equating a real number, v_0/ω_n, to an imaginary number, $-i(X_1 - X_2)$, but you have to look back at the assumed solution in Equation 3.53. When we assumed the exponential solutions, we never said that X_1 and X_2 were real amplitudes. They are, in fact, complex numbers. You can solve to find

$$X_1 = \frac{x_0}{2} + i\frac{v_0}{2\omega_n}$$

and

$$X_2 = \frac{x_0}{2} - i\frac{v_0}{2\omega_n}$$

so that both $(X_1 + X_2)$ and $-i(X_1 - X_2)$ are real amplitudes in Equation 3.55.

Once again employing Euler's Identity, we can write the response as

$$x(t) = e^{-\zeta\omega_n t}[X_1(\cos\omega_d t - i\sin\omega_d t) + X_2(\cos\omega_d t + i\sin\omega_d t)] \tag{3.63}$$

or

$$x(t) = e^{-\zeta\omega_n t}[(X_1 + X_2)\cos\omega_d t - i(X_1 - X_2)\sin\omega_d t] \tag{3.64}$$

Once again, we differentiate the displacement with respect to time in preparation for applying initial conditions, finding

$$\dot{x}(t) = -\zeta\omega_n e^{-\zeta\omega_n t}[(X_1 + X_2)\cos\omega_d t - i(X_1 - X_2)\sin\omega_d t]$$
$$+ e^{-\zeta\omega_n t}[-\omega_d(X_1 + X_2)\sin\omega_d t - i\omega_d(X_1 - X_2)\cos\omega_d t] \tag{3.65}$$

For $x(0) = x_0$, Equation 3.64 gives

$$x(0) = (X_1 + X_2)\cos 0 - i(X_1 - X_2)\sin 0 = x_0 \tag{3.66}$$

where the exponential function with a zero exponent has been set equal to one and from which we find

$$(X_1 + X_2) = x_0 \tag{3.67}$$

For $\dot{x}(0) = v_0$, Equation 3.65 yields

$$\dot{x}(0) = -\zeta\omega_n[(X_1 + X_2)\cos 0 - i(X_1 - X_2)\sin 0]$$
$$+ [-\omega_d(X_1 + X_2)\sin 0 - i\omega_d(X_1 - X_2)\cos 0] = v_0 \tag{3.68}$$

or

$$-\zeta\omega_n(X_1 + X_2) - i\omega_d(X_1 - X_2) = v_0 \tag{3.69}$$

Substituting for $(X_1 + X_2)$ from Equation 3.67 gives

$$-\zeta\omega_n x_0 - i\omega_d(X_1 - X_2) = v_0 \tag{3.70}$$

from which

$$-i(X_1 - X_2) = \frac{v_0 + \zeta\omega_n x_0}{\omega_d} \tag{3.71}$$

Finally, the results of Equations 3.67 and 3.71 can be substituted into Equation 3.64 to get an expression for the response of an underdamped system to initial conditions

$$x(t) = e^{-\zeta\omega_n t}\left[x_0\cos\omega_d t + \frac{v_0 + \zeta\omega_n x_0}{\omega_d}\sin\omega_d t\right] \tag{3.72}$$

Figure 3.5 shows a typical underdamped response. The oscillation is at the damped natural frequency so that the period, T, is defined by

$$\omega_d T = 2\pi \text{ so that } T = \frac{2\pi}{\omega_d} = \frac{2\pi}{\omega_n\sqrt{1-\zeta^2}} \tag{3.73}$$

The amplitude decreases exponentially with time. This is characterized by the two dashed lines in Figure 3.5 which are actually plots of the term

$$e^{-\zeta\omega_n t}$$

that modulates the response in Equation 3.72. The two peak responses marked as X_i and X_j will be used later when we discuss logarithmic decrement.

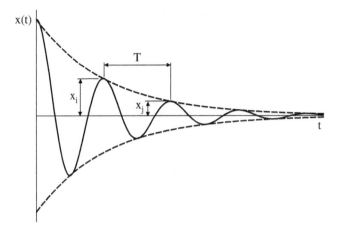

Figure 3.5 The underdamped response.

3.3.4 Critically Damped

The system is critically damped when $\zeta = 1$. The roots from Equation 3.52 are negative real and equal

$$\lambda_1 = -\omega_n \; ; \; \lambda_2 = -\omega_n$$

with the solution being

$$x(t) = (X_1 + X_2 t)e^{-\omega_n t} \tag{3.74}$$

Differentiating yields the velocity expression

$$\dot{x}(t) = X_2 e^{-\omega_n t} - \omega_n (X_1 + X_2 t)e^{-\omega_n t} \tag{3.75}$$

Applying the initial conditions gives, for $x(0) = x_0$

$$x(0) = X_1 = x_0 \tag{3.76}$$

and, for $\dot{x}(0) = v_0$

$$\dot{x}(0) = X_2 - \omega_n X_1 = v_0 \tag{3.77}$$

These give

$$X_1 = x_0$$

and

$$X_2 = v_0 + \omega_n x_0$$

so that the displacement can be written as

$$x(t) = [x_0 + (v_0 + \omega_n x_0)t]e^{-\omega_n t} \tag{3.78}$$

Figure 3.6 shows the response of the critically damped system for three different initial velocities. Critical damping is the lowest level of damping that suppresses oscillations. It also gives the fastest return to equilibrium of any level of damping. Underdamped systems

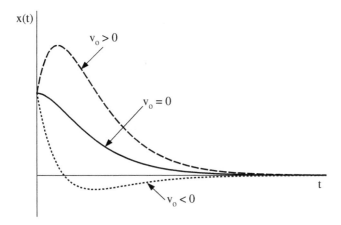

Figure 3.6 The critically damped response.

tend to oscillate around equilibrium for extended periods and overdamped systems have a slow response due to their high level of damping. Note from the figure that a large enough negative initial velocity can cause the critically damped system to have one, and only one, overshoot before returning exponentially to the equilibrium position.

3.3.5 Overdamped

For $\zeta > 1$, the system is overdamped and the roots from Equation 3.52 are

$$\lambda_1 = -\zeta\omega_n - \omega_n\sqrt{\zeta^2 - 1} \; ; \; \lambda_2 = -\zeta\omega_n + \omega_n\sqrt{\zeta^2 - 1}$$

with the superimposed exponential solution being

$$x(t) = X_1 e^{(-\zeta\omega_n - \omega_n\sqrt{\zeta^2-1})t} + X_2 e^{(-\zeta\omega_n + \omega_n\sqrt{\zeta^2-1})t} \tag{3.79}$$

which can be written as

$$x(t) = e^{-\zeta\omega_n t}(X_1 e^{-\omega_n\sqrt{\zeta^2-1}t} + X_2 e^{+\omega_n\sqrt{\zeta^2-1}t}) \tag{3.80}$$

Applying the initial conditions, without elaboration this time, leads to

$$X_1 = \frac{-v_0 + \omega_n x_0(-\zeta + \sqrt{\zeta^2 - 1})}{2\omega_n\sqrt{\zeta^2 - 1}}$$

and

$$X_2 = \frac{v_0 + \omega_n x_0(\zeta + \sqrt{\zeta^2 - 1})}{2\omega_n\sqrt{\zeta^2 - 1}}$$

Figure 3.7 shows a typical overdamped response. The total response is shown by the solid line with the contributions of the two roots, λ_1 and λ_2, shown by the dashed lines. λ_1 is larger in magnitude than λ_2 so the decaying exponential corresponding to λ_1 disappears rapidly compared to that associated with λ_2 which lies quite near the origin. Overall, the response is dominated by the root having the smallest magnitude. The closer that root lies to the origin, the slower the return to equilibrium will be.

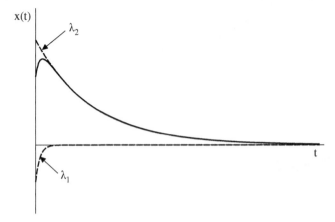

Figure 3.7 The overdamped response.

3.4 Root Locus

It is interesting to look at the motion of the roots of the damped system on the complex plane as damping is increased from zero, through critical damping, and into the overdamped region. Figure 3.8 shows this progression. The roots are, as stated in Equation 3.52

$$\lambda = -\zeta \omega_n \pm \omega_n \sqrt{\zeta^2 - 1} \qquad (3.81)$$

Figure 3.8 The root locus.

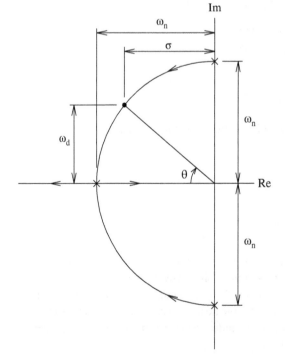

When $\zeta = 0$, the roots are $\pm i\omega_n$ and lie on the imaginary axis as shown in the figure.

As damping is increased, the roots move to the left and closer to the real axis. These roots have both real and imaginary components and are therefore in the underdamped region. The figure shows an underdamped root at $-\sigma + i\omega_d$. Since this is the underdamped case, we rewrite Equation 3.81 with $\zeta < 1$ to get

$$\lambda = -\zeta\omega_n \pm i\omega_n \sqrt{1 - \zeta^2} \tag{3.82}$$

and then compare the roots to find

$$\sigma = \zeta\omega_n$$

and

$$\omega_d = \omega_n \sqrt{1 - \zeta^2}$$

Most interesting about this result is that the length of the line connecting the root to the origin in the figure can be calculated, using Pythagoras' theorem, to be

$$\sqrt{\sigma^2 + \omega_d^2} = \sqrt{\zeta^2\omega_n^2 + \omega_n^2(1 - \zeta^2)} = \omega_n$$

This means that the trajectory of the roots from the undamped case to the critically damped case where they both lie at the same point on the real axis is a circular arc. The useful thing about this is that we can look at the cosine of the angle θ and write

$$\cos\theta = \frac{\sigma}{\omega_n} = \frac{\zeta\omega_n}{\omega_n} = \zeta$$

In other words, given a general complex root $\lambda = -\sigma + i\omega_d$, we can immediately say that its damped natural frequency is the magnitude of the imaginary part and its damping ratio is

$$\zeta = \cos\theta = \frac{\sigma}{\sqrt{\sigma^2 + \omega_d^2}}$$

Exercises

3.1 For the system shown in Figure E3.1, the pulleys are massless and frictionless and the cable is massless and inextensible.

(a) Find the equilibrium tension in the cable.

(b) Find an expression for c that gives the system critical damping.

3.2 The large uniform rectangular body (mass $= 15m$; width $= 2d$; height $= 4d$) rotates freely in the horizontal plane about a vertical axis through O. Its motion is resisted by the spring (stiffness $= k$) and the damper (damping coefficient $= c$). The spring and the damper are perpendicular to the edges of the block in the equilibrium position shown.

(a) Use Newton's laws to derive the equation-of-motion for small oscillations around equilibrium.

(b) Repeat part (a) using Lagrange's Equation and ensure that the results are the same.

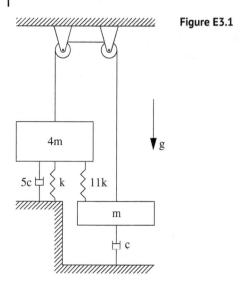

Figure E3.1

(c) Write the equation of motion in standard form and find expressions for the undamped natural frequency (ω_n) and the damping ratio (ζ).

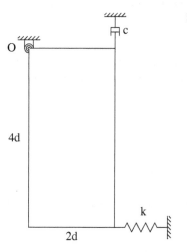

Figure E3.2

3.3 The rigid body is a uniform slender rod of length $6d$ having a mass m. It is shown in its equilibrium position and is able to pivot without friction about point A. Small rotations about A are resisted by the spring and damper shown.
  (a) Use Newton's laws to derive the equation-of-motion for small rotations around equilibrium.
  (b) Repeat part (a) using Lagrange's Equation and ensure that the results are the same.

(c) Write the equation of motion in standard form and find expressions for the undamped natural frequency (ω_n) and the damping ratio (ζ).

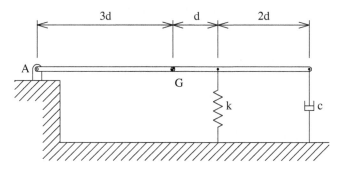

Figure E3.3

3.4 Figure E3.4 shows the equilibrium positions of three orientations of a body made up of a point mass, m, supported at the end of a massless, rigid, rod that is free to pivot about a frictionless pin connecting it to ground. There is a single spring of stiffness, k, at the midpoint of the rod, that resists motions away from equilibrium.

(a) Derive the equation of motion for each of the three orientations, put them in standard form, and find the natural frequencies. Note that gravitational forces must be included if they are restoring forces, as is the case for two of these orientations.

(b) Let the system be inclined at an arbitrary angle, ϕ, measured counterclockwise from the horizontal orientation. Derive the equation of motion and find a general expression for the natural frequency as a function of ϕ.

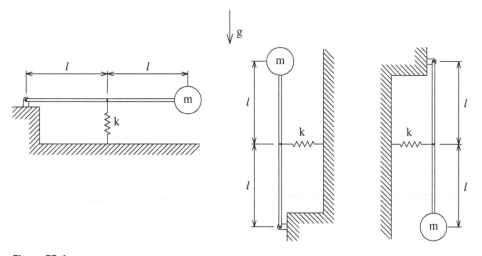

Figure E3.4

3.5 Two concentric disks (radii of r and $2r$) are fixed together to form a single rigid body that rotates freely in a vertical plane about a pin that connects them to ground at their mutual center point, O. The mass of this rigid body is $3m$ and it has a radius of gyration, k_O, equal to r. There is a horizontal spring of stiffness, k, that connects the body to the ground a distance, r, above the center point. There is also a mass, m, that is suspended from the outer edge of the concentric disks using a massless, inextensible cord. What is the natural frequency of this system?

Figure E3.5

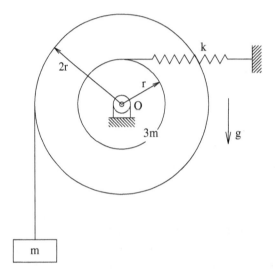

3.6 Derive expressions for the undamped natural frequency and the damping ratio for the system shown. The pulleys are massless and frictionless and the cord is inextensible.

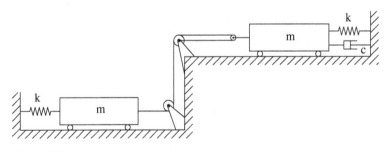

Figure E3.6

3.7 Derive expressions for the undamped natural frequency and the damping ratio for the system shown. Treat the T-shaped structure as being rigid and massless. The mass, m, is a particle.

Figure E3.7

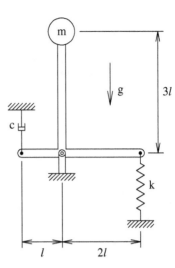

4

SDOF Systems – Forced Vibrations – Response to Initial Conditions

We move on to considering the response of undamped single degree of freedom (SDOF) systems as they respond to harmonic applied forces and initial conditions.

4.1 Time Response to a Harmonically Applied Force in Undamped Systems

Figure 4.1 shows a single degree of freedom undamped system subjected to a harmonically varying force

$$F(t) = F \cos \omega t$$

with magnitude, F, and *forcing frequency*, ω. It is important to realize that the forcing frequency is a specified parameter and is not the natural frequency of the system.

Figure 4.2 shows a free body diagram of the system. The force balance equation is

$$+\uparrow \sum F : F(t) - kx = m\ddot{x} \tag{4.1}$$

from which the equation of motion is found to be

$$m\ddot{x} + kx = F(t) \tag{4.2}$$

or, substituting the expression for the applied force,

$$m\ddot{x} + kx = F \cos \omega t \tag{4.3}$$

We need to consider the solutions to Equation 4.3. These solutions will tell us how the system responds to the applied force.

We previously considered free vibration problems where the differential equations had zero on the right-hand side. Equations of that type are called *homogeneous* and the systems that they represent exhibit their natural response to initial conditions. Forced vibration problems have a non-zero right-hand side so they are *non-homogeneous* and there are two parts to their response. The system is still the same system and has the same natural frequency as it did before so there will be part of the response that is derived from the homogeneous system. This is called the *complementary solution*, $x_c(t)$. For the system being considered here, we know that the natural response is an undamped oscillation at the natural frequency, $\omega_n = \sqrt{k/m}$ and we can write that as

$$x_c(t) = A \cos \omega_n t + B \sin \omega_n t \tag{4.4}$$

Introduction to Mechanical Vibrations, First Edition. Ronald J. Anderson.
© 2020 John Wiley & Sons Ltd. Published 2020 by John Wiley & Sons Ltd.
Companion website: www.wiley.com/go/anderson/introduction-to-vibrations

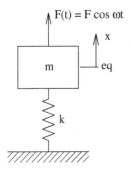

Figure 4.1 SDOF system with a harmonically applied force.

$F(t) = F \cos \omega t$

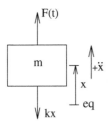

Figure 4.2 SDOF system with a harmonically applied force – FBD.

where A and B are to be determined from the initial conditions.

The second part to the solution of non-homogeneous differential equations comes from the fact that the system is being forced to respond in such a way that the left-hand side of Equation 4.3 is exactly equal to the right-hand side at any time, t. It is clear that this can only happen if the response is an oscillation at the forcing frequency, ω. In fact, both $x(t)$ and $\ddot{x}(t)$ need to have $\cos \omega t$ in them or the equation will never be satisfied. This solution is called the particular solution, $x_p(t)$ and is

$$x_p(t) = X \cos \omega t \tag{4.5}$$

Differentiating Equation 4.5 twice yields

$$\ddot{x}_p(t) = -\omega^2 X \cos \omega t \tag{4.6}$$

and substituting the expressions from Equations 4.5 and 4.6 into Equation 4.3 gives

$$(-m\omega^2 + k)X \cos \omega t = F \cos \omega t \tag{4.7}$$

from which we find

$$X = \frac{F}{(k - m\omega^2)} \tag{4.8}$$

so that the particular solution is

$$x_p(t) = \frac{F}{(k - m\omega^2)} \cos \omega t \tag{4.9}$$

The total solution is the superposition of the complementary and particular solutions. That is

$$x(t) = x_c(t) + x_p(t) \tag{4.10}$$

Figure 4.3   Widely separated frequencies.

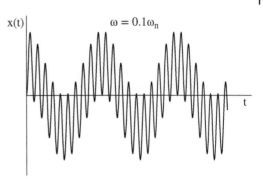

or

$$x(t) = A \cos \omega_n t + B \sin \omega_n t + \frac{F}{(k - m\omega^2)} \cos \omega t \qquad (4.11)$$

where A and B still have to be determined from initial conditions.

Let the initial displacement be x_0 and the initial velocity be v_0. We differentiate Equation 4.11 to find

$$\dot{x}(t) = -A\omega_n \sin \omega_n t + B\omega_n \cos \omega_n t - \frac{F\omega}{(k - m\omega^2)} \sin \omega t \qquad (4.12)$$

and then evaluate Equations 4.11 and 4.12 at time $t = 0$ to find

$$x_0 = A + \frac{F}{(k - m\omega^2)} \qquad (4.13)$$

and

$$v_0 = B\omega_n \qquad (4.14)$$

so that we can write the displacement as

$$x(t) = \left[x_0 - \frac{F}{(k - m\omega^2)} \right] \cos \omega_n t + \left[\frac{v_0}{\omega_n} \right] \sin \omega_n t + \frac{F}{(k - m\omega^2)} \cos \omega t \qquad (4.15)$$

Figures 4.3 to 4.5 show how the system reacts to a harmonic force applied at time $t = 0$. The results shown in the figures are calculated with zero initial conditions for convenience. That is, both x_0 and v_0 have been set to zero so we are simply seeing the system respond to the sudden application of the harmonically varying force.

Figure 4.3 shows the response in the case where the forcing frequency and the natural frequency are relatively far apart. The result is two superimposed waves of different frequency that result in a fairly regular oscillation.

There are two phenomena of particular interest. These are *beating* and *resonance*.

4.1.1   Beating

Figure 4.4 shows the case where the natural frequency and the forcing frequency are very close to each other, resulting in the phenomenon called *beating*. The behavior is as if there is a high-frequency wave whose amplitude is modulated by a low frequency wave. This can be explained as follows.

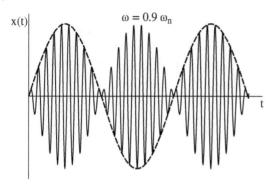

$\omega = 0.9\,\omega_n$

Figure 4.4   The *Beating* phenomenon.

We start with Equation 4.15 with $x_0 = 0$ and $v_0 = 0$

$$x(t) = \left[-\frac{F}{(k - m\omega^2)} \right] \cos \omega_n t + \left[\frac{F}{(k - m\omega^2)} \right] \cos \omega t \tag{4.16}$$

which can be rearranged to become

$$x(t) = \left[\frac{F}{(k - m\omega^2)} \right] (\cos \omega t - \cos \omega_n t) \tag{4.17}$$

We now define two new frequencies. The first is the average of the natural frequency and the forcing frequency, $\bar{\omega}$,

$$\bar{\omega} = \frac{\omega_n + \omega}{2} \tag{4.18}$$

and the second is half the difference between the natural frequency and the forcing frequency, $\Delta\omega$,

$$\Delta\omega = \frac{\omega_n - \omega}{2} \tag{4.19}$$

If we add $\bar{\omega}$ to $\Delta\omega$, we have

$$\bar{\omega} + \Delta\omega = \left(\frac{\omega_n}{2} + \frac{\omega}{2} \right) + \left(\frac{\omega_n}{2} - \frac{\omega}{2} \right) = \omega_n \tag{4.20}$$

and, if we subtract them, we find

$$\bar{\omega} - \Delta\omega = \left(\frac{\omega_n}{2} + \frac{\omega}{2} \right) - \left(\frac{\omega_n}{2} - \frac{\omega}{2} \right) = \omega \tag{4.21}$$

Using the results of Equations 4.20 and 4.21 allows the term

$$(\cos \omega t - \cos \omega_n t)$$

in Equation 4.17 to be written as

$$\cos \omega t - \cos \omega_n t = \cos(\bar{\omega} t - \Delta\omega t) - \cos(\bar{\omega} t + \Delta\omega t) \tag{4.22}$$

At this point we use the equation for the cosine of the sum of two angles

$$\cos(\theta + \phi) = \cos \theta \cos \phi - \sin \theta \sin \phi$$

to write

$$\cos(\bar{\omega} t - \Delta\omega t) = \cos \bar{\omega} t \cos \Delta\omega t + \sin \bar{\omega} t \sin \Delta\omega t \tag{4.23}$$

and

$$\cos(\overline{\omega}t + \Delta\omega t) = \cos\overline{\omega}t \cos\Delta\omega t - \sin\overline{\omega}t \sin\Delta\omega t \qquad (4.24)$$

Subtracting Equation 4.24 from Equation 4.23 yields

$$\cos(\overline{\omega}t - \Delta\omega t) - \cos(\overline{\omega}t + \Delta\omega t) = 2\sin\overline{\omega}t \sin\Delta\omega t \qquad (4.25)$$

which, substituted into Equation 4.22, gives

$$\cos\omega t - \cos\omega_n t = 2\sin\overline{\omega}t \sin\Delta\omega t \qquad (4.26)$$

Finally, we substitute the result from Equation 4.26 into Equation 4.17 and write the result as

$$x(t) = \left[\frac{2F}{(k - m\omega^2)} \sin\Delta\omega t \right] \sin\overline{\omega}t \qquad (4.27)$$

which can be viewed as an oscillation at the frequency $\overline{\omega}$ with an amplitude that varies harmonically at the frequency $\Delta\omega$. $\overline{\omega}$ is the average of the natural frequency and the forcing frequency and is approximately equal to either of them since they must be nearly the same for beating to occur. $\overline{\omega}$ is much larger than $\Delta\omega$, which is defined to be half the difference between the natural frequency and the forcing frequency.

An unmodulated wave with frequency $\overline{\omega}$ would have a constant amplitude, X, and could be written as

$$x(t) = X\sin\overline{\omega}t \qquad (4.28)$$

Comparing this to Equation 4.27, we can see that the amplitude for the beating system is

$$X = \left[\frac{2F}{(k - m\omega^2)} \sin\Delta\omega t \right] \qquad (4.29)$$

The amplitude therefore varies harmonically at the frequency $\Delta\omega$ giving the beating seen in Figure 4.4.

The period of a periodic signal is the length of time it takes for the signal to repeat itself. The dashed line shown in Figure 4.4 is a plot of the amplitude from Equation 4.29 and shows how the amplitude varies with time but is somewhat deceptive regarding the period of the beat. The period of the dashed line is $2\pi/\Delta\omega$ but close observation will show that the signal repeats itself twice during that time. The period of the beat is therefore $T_{beat} = \pi/\Delta\omega$ or, substituting from Equation 4.19,

$$T_{beat} = \frac{2\pi}{\omega_n - \omega} \qquad (4.30)$$

4.1.2 Resonance

When the forcing frequency becomes exactly equal to the natural frequency, the system goes into *resonance*. The solution to the equation of motion needs to change because the particular solution given in Equation 4.5 is now exactly the same as part of the complementary solution. That is, the complementary solution is still

$$x_c(t) = A\cos\omega_n t + B\sin\omega_n t \qquad (4.31)$$

but the particular solution, which was

$$x_p(t) = X \cos \omega t \qquad (4.32)$$

is the same as $A \cos \omega_n t$ since $\omega = \omega_n$.

We need the particular solution to be linearly independent of the complementary solution. As a result, the particular solution we use is

$$x_p(t) = Xt \sin \omega_n t \qquad (4.33)$$

We differentiate $x_p(t)$ twice to get

$$\ddot{x}_p(t) = 2\omega_n X \cos \omega_n t - \omega_n^2 Xt \sin \omega_n t \qquad (4.34)$$

and substitute $x_p(t)$, $\ddot{x}_p(t)$, and $\omega = \omega_n$ into the equation of motion (Equation 4.3) to find

$$2m\omega_n X \cos \omega_n t - m\omega_n^2 Xt \sin \omega_n t + kXt \sin \omega_n t = F \cos \omega_n t \qquad (4.35)$$

which, upon noting that $m\omega_n^2 = k$, yields

$$X = \frac{F}{2m\omega_n} \qquad (4.36)$$

Then, superimposing the complementary and particular solutions, $x(t)$ becomes

$$x(t) = A \cos \omega_n t + B \sin \omega_n t + \frac{F}{2m\omega_n} t \sin \omega_n t \qquad (4.37)$$

and it is a simple matter to show that $A = x_0$ and $B = v_0/\omega_n$ so that

$$x(t) = x_0 \cos \omega_n t + \frac{v_0}{\omega_n} \sin \omega_n t + \frac{F}{2m\omega_n} t \sin \omega_n t \qquad (4.38)$$

Equation 4.38 shows that the response includes a term that grows linearly with time. The amplitude therefore grows without bound as shown in Figure 4.5 and the amplitude will theoretically become infinite.

This is a good place to remind ourselves that the models we are using are linear and valid only for small motions away from equilibrium. In fact, the amplitudes don't become infinite because they are limited by nonlinearities and the damping that is present in all systems but it also a fact that resonance conditions exist and can do serious damage to systems even if the amplitudes are finite.

Figure 4.5 Resonance.

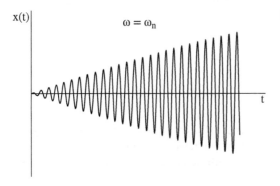

Exercises

4.1   An airplane wing with a natural frequency in bending of 40 Hz is driven harmonically by the rotation of its engines at 39.9 Hz. What is the period of the resulting beat?

4.2   The figure shows a high-frequency acceleration signal which shows evidence of beating. Estimate the difference between the natural frequency and the forcing frequency, in Hz, for these data.

Figure E4.2

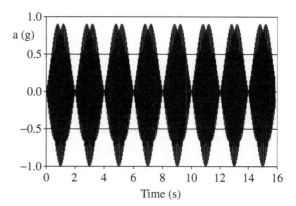

4.3   North American highways often have rumble strips on the shoulders to cause tire noise that warns drivers that they have drifted outside of their lane. A typical rumble strip is approximately sinusoidal with a wavelength of 1 ft (0.30 m) and an amplitude of 0.5 in (1.3 cm). Given that you are driving a car for which the bounce mode has a natural frequency of 1.5 Hz, at what speed would you experience resonance?

5

SDOF Systems – Steady State Forced Vibrations

Chapter 4 looked at the response of undamped SDOF systems to harmonic forces and initial conditions. Here we look at the response of the forced system after the forcing function has been applied for a long time. Every real system has some level of inherent damping so that the complementary solution to the equation of motion will eventually die away. The complementary solution is therefore called the *transient solution*. The magnitude of the particular solution is affected by damping but the response will persist so long as the forcing function is present. This sustained response is called the *steady-state solution*.

5.1 Undamped Steady State Response to a Harmonically Applied Force

Considering first the undamped SDOF case, Figure 5.1 shows the system being considered. Recall from the previous section that the equation of motion is

$$m\ddot{x} + kx = F\cos\omega t \tag{5.1}$$

where the applied force has been taken to be

$$F(t) = F\cos\omega t \tag{5.2}$$

The particular solution to Equation 5.1 must be of the form

$$x_p(t) = X\cos\omega t \tag{5.3}$$

We differentiate $x_p(t)$ twice and substitute into the equation of motion to find

$$(-m\omega^2 + k)X\cos\omega t = F\cos\omega t \tag{5.4}$$

from which the steady state amplitude is determined to be

$$X = \frac{F}{(k - m\omega^2)} \tag{5.5}$$

We are interested in the amplitude of the response as a function of forcing frequency. Equation 5.5 doesn't impart much of a feeling about how the amplitude varies with frequency but, with a little manipulation, we can make it more understandable. We start

Introduction to Mechanical Vibrations, First Edition. Ronald J. Anderson.
© 2020 John Wiley & Sons Ltd. Published 2020 by John Wiley & Sons Ltd.
Companion website: www.wiley.com/go/anderson/introduction-to-vibrations

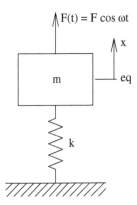

$F(t) = F \cos \omega t$

m

eq

x

k

Figure 5.1   SDOF system with a harmonically applied force.

by dividing both the numerator and the denominator by k to get

$$X = \frac{F/k}{\left(1 - \frac{m\omega^2}{k}\right)} \tag{5.6}$$

Then we recognize that F/k is actually the motion that would result if a force having the same magnitude as the harmonic force were to be applied statically to the system. That is, if the force balance on the mass were written for the static case where $F(t) = F = $ a constant and the acceleration term doesn't exist because there is no motion, the result would be

$$+\uparrow \sum F : F - kX_0 = 0 \tag{5.7}$$

where X_0 is defined to be the static displacement. Clearly the result is that $X_0 = F/k$ so that Equation 5.6 becomes

$$X = \frac{X_0}{\left(1 - \frac{m\omega^2}{k}\right)} \tag{5.8}$$

Next we replace the m/k term in the denominator with $1/\omega_n^2$ to get

$$X = \frac{X_0}{1 - \left(\frac{\omega}{\omega_n}\right)^2} \tag{5.9}$$

and, finally, we write a dimensionless expression for the magnitude of the dynamic forced response divided by the magnitude of the static forced response for equal force magnitudes

$$\frac{X}{X_0} = \frac{1}{1 - \left(\frac{\omega}{\omega_n}\right)^2} \tag{5.10}$$

The ratio X/X_0 is plotted against the frequency ratio, ω/ω_n, in Figure 5.2. There are several things to note about the frequency response of the undamped SDOF system.

- At zero frequency, $X/X_0 = 1$. Zero frequency corresponds to the static case where the force is not oscillating so this result is entirely expected.

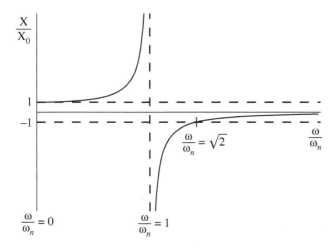

Figure 5.2 Frequency response – SDOF undamped system.

- X/X_0 increases as the forcing frequency increases and eventually goes to $+\infty$ as the frequency ratio becomes one (i.e. when $\omega = \omega_n$). The predicted infinite response is strictly due to the assumption of linear behavior of the spring in the system. In fact, nonlinearities (things like coils in a coil spring bumping into each other for instance) limit the response to finite values. The response is large but not infinite.

- X/X_0 goes from $+\infty$ to $-\infty$ as the frequency ratio passes through one. This indicates a phase shift. When the frequency ratio is less than one, X/X_0 is positive, indicating that both X and X_0 have the same sign. This means they are in phase with each other and, in fact, in phase with the applied force since X_0 and F are always in phase. On the other hand, X/X_0 is negative when the frequency ratio is greater than one indicating that they are out of phase with each other for forcing frequencies above the natural frequency. If you consider for a moment what it means to be out of phase, you will get a feeling for why the amplitudes tend to zero for very high forcing frequencies. X being out of phase with F means that when the mass is displaced upwards the force is pushing downwards and vice versa. Essentially, the fact that the force is opposing the motion causes all motion to stop when the frequency becomes high enough.

- The magnitude of the dynamic displacement, X, is equal to the magnitude of the static displacement, X_0, at only two frequency ratios – zero and $\sqrt{2}$. The dynamic magnitude is less than the static magnitude only for frequency ratios greater than $\sqrt{2}$.

Figure 5.3 shows the absolute value of X/X_0 plotted against the frequency ratio. $|X/X_0|$ gives amplitude information without reference to the phase relationship and is called the *Magnification Factor* or simply the *Amplitude Ratio*. We will denote it by the variable M standing for Magnification Factor. Reference to plots of the Magnification Factor are the most common way of looking at the frequency response of forced systems.

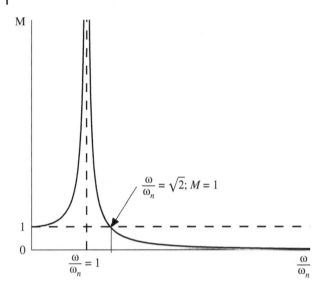

Figure 5.3 Magnification Factor - SDOF undamped system.

5.2 Damped Steady State Response to a Harmonically Applied Force

We now consider adding damping to the system. Figure 5.4 shows the system. The equation of motion is

$$m\ddot{x} + c\dot{x} + kx = F(t) \tag{5.11}$$

We need to look at all possible forms for a harmonic forcing function keeping in mind that we will need to find the particular solution for Equation 5.11. Note that we need to add together a function and its first two derivatives on the left-hand side and make the sum equal to a single function on the right-hand side. This is clearly a case where the complex exponential function is the best choice[1].

Therefore, let

$$F(t) = Fe^{i\omega t} \tag{5.12}$$

Before proceeding any farther, it is advisable to write the equation of motion in the standard form. We therefore divide Equation 5.11 by m and write

$$\ddot{x} + 2\zeta\omega_n\dot{x} + \omega_n^2 x = \frac{F}{m}e^{i\omega t} \tag{5.13}$$

---

1 We could put $F\cos\omega t$ on the right-hand side and then use

$$x(t) = A\cos\omega t + B\sin\omega t$$

as the particular solution but the level of effort required would be much higher. We would find two simultaneous equations from which we could find A and B. Then we would have to combine A and B to get an amplitude and a phase angle. It's much simpler to use the complex exponential with a single amplitude.

Figure 5.4 Damped SDOF system with a harmonically applied force.

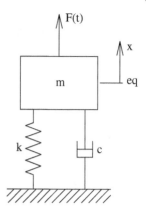

and assume that the steady state solution for $x(t)$ is

$$x(t) = Xe^{i\omega t} \tag{5.14}$$

with derivatives

$$\dot{x}(t) = i\omega Xe^{i\omega t} \tag{5.15}$$

and

$$\ddot{x}(t) = i\omega(i\omega Xe^{i\omega t}) = -\omega^2 Xe^{i\omega t} \tag{5.16}$$

Substituting into Equation 5.13 gives

$$[-\omega^2 + i(2\zeta\omega_n\omega) + \omega_n^2]Xe^{i\omega t} = \frac{F}{m}e^{i\omega t} \tag{5.17}$$

from which we can solve directly for X which, after a little rearranging of terms, becomes

$$X = \frac{\frac{F}{m}}{[(\omega_n^2 - \omega^2) + i(2\zeta\omega_n\omega)]} \tag{5.18}$$

Admittedly, X in Equation 5.18 is a complex number, but that simply indicates that the response can be written as an amplitude and a phase angle. Before working out the amplitude and phase angle we will manipulate Equation 5.18 in order to make it dimensionless. We start by dividing the numerator and denominator by ω_n^2 to get

$$X = \frac{\frac{F}{m\omega_n^2}}{\left[\left\{1 - \left(\frac{\omega}{\omega_n}\right)^2\right\} + i\left(2\zeta\frac{\omega}{\omega_n}\right)\right]} \tag{5.19}$$

Then we note that the term, $m\omega_n^2$, in the numerator is simply equal to k so that the numerator becomes F/k which is the static deflection X_0 that we found in Equation 5.7. We can then write

$$\frac{X}{X_0} = \frac{1}{\left[\left\{1 - \left(\frac{\omega}{\omega_n}\right)^2\right\} + i\left(2\zeta\frac{\omega}{\omega_n}\right)\right]} \tag{5.20}$$

This is a good time for a short review of complex numbers. Any complex number $a + ib$, can be represented by a magnitude and a phase angle. If you plot $a + ib$ as a point on the complex plane you will see that you can represent the same point in polar coordinates with a line of length $\sqrt{a^2 + b^2}$ that lies at an angle ϕ from the real axis where $\tan \phi = b/a$. The number can equally well be represented as a complex exponential using Euler's Identity. That is, we can write

$$a + ib = \sqrt{a^2 + b^2} e^{i\phi} = \sqrt{a^2 + b^2} (\cos \phi + i \sin \phi)$$

For what we are about to do, the exponential notation is preferable. For two complex numbers, $a + ib$ and $c + id$, we can define two amplitudes, $\sqrt{a^2 + b^2}$ and $\sqrt{c^2 + d^2}$, and two phase angles, $\phi_1 = \tan^{-1}(b/a)$ and $\phi_2 = \tan^{-1}(d/c)$, and then write their ratio as

$$\frac{a + ib}{c + id} = \frac{\sqrt{a^2 + b^2} e^{i\phi_1}}{\sqrt{c^2 + d^2} e^{i\phi_2}} = \frac{\sqrt{a^2 + b^2}}{\sqrt{c^2 + d^2}} e^{i(\phi_1 - \phi_2)}$$

The result is that the ratio of two complex numbers has a magnitude that is simply the ratio of the individual magnitudes and a phase angle that is the difference between the two phase angles. Applying this to Equation 5.20 gives the magnification factor as

$$M = \left| \frac{X}{X_0} \right| = \frac{1}{\sqrt{\left\{ 1 - \left(\frac{\omega}{\omega_n} \right)^2 \right\}^2 + \left(2\zeta \frac{\omega}{\omega_n} \right)^2}} \tag{5.21}$$

with the phase angle

$$\phi = \tan^{-1} \left(\frac{-2\zeta \frac{\omega}{\omega_n}}{1 - \left(\frac{\omega}{\omega_n} \right)^2} \right) \tag{5.22}$$

Figure 5.5 shows the magnification factor plotted versus frequency ratio for different damping ratios. Things to notice about the figure are:

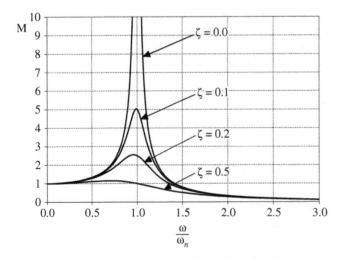

Figure 5.5   Magnification Factor – SDOF damped system.

- The infinite response predicted with zero damping is very much attenuated, even with relatively low levels of damping.
- The peak response occurs at values of ω/ω_n slightly less than one. The actual peak is at

$$\omega/\omega_n = \sqrt{1 - 2\zeta^2}$$

- The peak magnification can be approximated by letting $(\omega/\omega_n) = 1$ in Equation 5.21, giving

$$M_{max} \approx \frac{1}{2\zeta}$$

5.3 Response to Harmonic Base Motion

The analysis of the system with the harmonic applied force in the previous section brought out some properties of the dynamic response that are useful but the model itself (i.e. the idea that a force is applied directly to the mass) is somewhat unrealistic. There are few instances when this happens in practice.

A realistic model of forced vibrations is one where the system is responding to harmonic motions of the base to which it is mounted. Stationary equipment in power plants, for instance, is mounted on floors that are continuously moving at the frequency of operation of the very large generators. Even houses alongside busy highways are subjected to ground-borne vibrations from traffic passing by.

Our analysis of harmonic base motions (also known as seismic disturbances) starts with the system shown in Figure 5.6. The base motion, $y(t)$, is

$$y(t) = Ye^{i\omega t} \tag{5.23}$$

where Y is the amplitude and ω is the forcing frequency.

Figure 5.7 shows a free body diagram of the system.

Consider the force in the spring for a moment. Both ends of the spring are connected to moving bodies in this system unlike previous systems we have analyzed. The spring will only have a force in it if it changes in length. We call the change in length the *spring deflection*, δ and then assign a spring force, $F_s = k\delta$. It is relatively clear that the spring deflection in this case will require that x and y be different. That is, if $x = y$, the spring simply moves

Figure 5.6 Damped SDOF system with harmonic base motion.

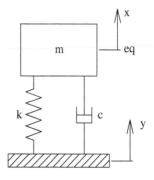

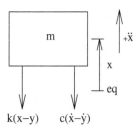

$k(x-y)$ $c(\dot{x}-\dot{y})$

in space but doesn't change in length. We can therefore say that $\delta = x - y$ or $\delta = y - x$. The question is *Which one is it?* The answer is that either can be used but there must be agreement with the direction of the force shown on the free body diagram. The spring force in Figure 5.7 is drawn as if the spring is in tension so that it is pulling on the mass. The spring can only be in tension if $x > y$ so that the top moves up farther than the bottom. If $x > y$, then $\delta = x - y$ is positive and it must be used as shown on the free body diagram. You could equally as well assume compression on the free body diagram and draw the arrows in the other direction but then you would need to recognize that compression requires $y > x$ so you would use $\delta = y - x$ for the positive compression deflection. A similar argument applies to the damper force.

The force balance equation is

$$+\uparrow \sum F : -k(x - y) - c(\dot{x} - \dot{y}) = m\ddot{x} \tag{5.24}$$

from which

$$m\ddot{x} + c\dot{x} + kx = c\dot{y} + ky \tag{5.25}$$

Equation 5.25 can be written in standard form by dividing throughout by m. The result is

$$\ddot{x} + 2\zeta\omega_n\dot{x} + \omega_n^2 x = 2\zeta\omega_n\dot{y} + \omega_n^2 y \tag{5.26}$$

which becomes, after substituting $y(t)$ from Equation 5.23,

$$\ddot{x} + 2\zeta\omega_n\dot{x} + \omega_n^2 x = (i2\zeta\omega_n\omega + \omega_n^2)Ye^{i\omega t} \tag{5.27}$$

The steady state solution will then be

$$x(t) = Xe^{i\omega t} \tag{5.28}$$

which, when differentiated and substituted into Equation 5.27, yields

$$[-\omega^2 + 2\zeta\omega_n\omega i + \omega_n^2]Xe^{i\omega t} = (2\zeta\omega_n\omega i + \omega_n^2)Ye^{i\omega t} \tag{5.29}$$

We perform manipulations similar to those in the previous section. First we solve for X

$$X = \frac{(2\zeta\omega_n\omega i + \omega_n^2)}{[-\omega^2 + 2\zeta\omega_n\omega i + \omega_n^2]}Y \tag{5.30}$$

then we reorganize terms a little and divide numerator and denominator by ω_n^2, resulting in

$$X = \frac{1 + \left(2\zeta\frac{\omega}{\omega_n}\right)i}{\left[1 - \left(\frac{\omega}{\omega_n}\right)^2\right] + \left(2\zeta\frac{\omega}{\omega_n}\right)i}Y \tag{5.31}$$

The ratio of the amplitude of motion of the mass, X, to the amplitude of motion of the base, Y is

$$\frac{X}{Y} = \frac{1 + \left(2\zeta\frac{\omega}{\omega_n}\right)i}{\left[1 - \left(\frac{\omega}{\omega_n}\right)^2\right] + \left(2\zeta\frac{\omega}{\omega_n}\right)i} \tag{5.32}$$

The magnitude of X/Y is defined as the Magnification Factor, M, and is

$$M = \sqrt{\frac{1 + \left(2\zeta\frac{\omega}{\omega_n}\right)^2}{\left[1 - \left(\frac{\omega}{\omega_n}\right)^2\right]^2 + \left(2\zeta\frac{\omega}{\omega_n}\right)^2}} \tag{5.33}$$

Figure 5.8 shows the Magnification Factor plotted versus frequency ratio for different damping ratios. This figure is very much like Figure 5.5 for directly applied forces. One major difference is that the Magnification Factor is equal to one when the frequency ratio is $\sqrt{2}$ for all damping ratios. We can show this by letting $r = \omega/\omega_n$ in Equation 5.33, then setting $M = 1$, and solving for r as follows.

$$1 = \sqrt{\frac{1 + (2\zeta r)^2}{(1 - r^2)^2 + (2\zeta r)^2}} \tag{5.34}$$

Square both sides to get

$$1 = \frac{1 + (2\zeta r)^2}{(1 - r^2)^2 + (2\zeta r)^2} \tag{5.35}$$

Multiply both sides by the denominator to get

$$(1 - r^2)^2 + (2\zeta r)^2 = 1 + (2\zeta r)^2 \tag{5.36}$$

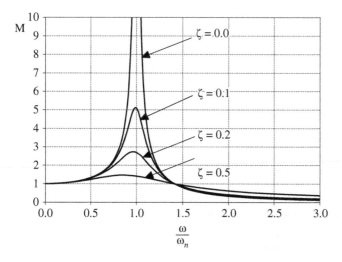

Figure 5.8 Magnification Factor – Harmonic base motion.

Note that $(2\zeta r)^2$ appears on both sides and cancel it out, leaving

$$(1 - r^2)^2 = 1 \tag{5.37}$$

from which

$$(1 - r^2) = \pm 1 \tag{5.38}$$

resulting in the solutions $r = 0$ and $r = \sqrt{2}$. These are in fact the two frequency ratios where the displacement of the mass is equal to the displacement of the base.

Another difference is that increasing the damping ratio increases the magnitude of the response for all frequency ratios above $\sqrt{2}$. This is somewhat counterintuitive since we expect damping to attenuate vibrations. The reason for it is that the damper is transmitting forces to the mass and, as the frequency increases, the velocity increases and the forces transmitted by the damper become dominant in the system.

When designing a system that will be operating in an environment where it will be subjected to base motion, it is very important to consider the magnitude of the forces being transmitted. We typically aim for the mass to have a comfortable ride in the sense that it should not experience large displacements. Equally as important is that the system mounting points should be able to accommodate the transmitted forces.

The free body diagram in Figure 5.7 shows that the total transmitted force, f_T, is the sum of the forces transmitted through the spring and the damper. That is

$$f_T = k(x - y) + c(\dot{x} - \dot{y}) \tag{5.39}$$

and, according to the force balance in Equation 5.24, this could also be written as

$$f_T = -m\ddot{x} \tag{5.40}$$

where we know that

$$x(t) = Xe^{i\omega t} \tag{5.41}$$

so that

$$\ddot{x}(t) = -\omega^2 Xe^{i\omega t} \tag{5.42}$$

so that the magnitude of the transmitted force, denoted by F_T, will be

$$F_T = m\omega^2 X \tag{5.43}$$

The Magnification Factor (Equation 5.33) can be used to express X as a function of Y, yielding

$$F_T = m\omega^2 Y \sqrt{\frac{1 + \left(2\zeta\frac{\omega}{\omega_n}\right)^2}{\left[1 - \left(\frac{\omega}{\omega_n}\right)^2\right]^2 + \left(2\zeta\frac{\omega}{\omega_n}\right)^2}} \tag{5.44}$$

and this expression can be multiplied and divided by ω_n^2 to get

$$F_T = m\omega_n^2 Y \left(\frac{\omega}{\omega_n}\right)^2 \sqrt{\frac{1 + \left(2\zeta\frac{\omega}{\omega_n}\right)^2}{\left[1 - \left(\frac{\omega}{\omega_n}\right)^2\right]^2 + \left(2\zeta\frac{\omega}{\omega_n}\right)^2}} \tag{5.45}$$

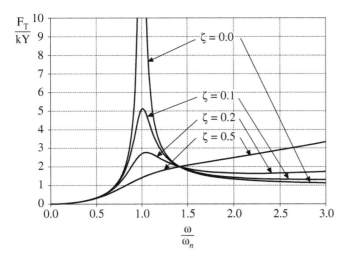

Figure 5.9 Force transmissibility – Harmonic base motion.

Finally we note that $m\omega_n^2 = k$ and write an expression for the *force transmissibility* which is defined to be the ratio of the transmitted force to the force that would be transmitted through the spring if its deflection was equal to the amplitude of base motion, Y. That is

$$\frac{F_T}{kY} = \left(\frac{\omega}{\omega_n}\right)^2 \sqrt{\frac{1 + \left(2\zeta\frac{\omega}{\omega_n}\right)^2}{\left[1 - \left(\frac{\omega}{\omega_n}\right)^2\right]^2 + \left(2\zeta\frac{\omega}{\omega_n}\right)^2}}$$ (5.46)

Figure 5.9 shows the force transmissibility for a selection of damping ratios as a function of frequency ratio. Note that all of the curves pass through the point ($F_T/kY = 2$, $\omega/\omega_n = \sqrt{2}$) and that the transmitted forces increase dramatically with increasing damping ratio for $\omega/\omega_n > \sqrt{2}$. Note also that increased damping ratio significantly decreases the force transmissibility in the range $0 < \omega/\omega_n < \sqrt{2}$.

Design of vibration isolation systems must take into account both the amplitude of the motions and the magnitude of the transmitted force. As is so often the case in design, there will inevitably be a tradeoff between the two.

5.4 Response to a Rotating Unbalance

Machines with rotating components often experience harmonic forced vibration. The force is due to an unbalance in the rotating mass caused by its center of mass not being on the axis of rotation. Figure 5.10 shows the model for the system. The total mass of the machine is m and the rotating mass component is m_r. The distance from the center of rotation, O, to the center of mass of the rotating component is the *eccentricity*, e. The constant rotational speed of the shaft is ω and the center of mass is shown in a position where it has rotated an

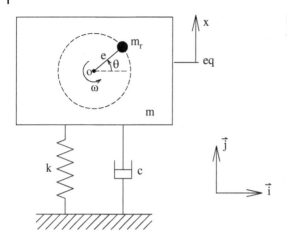

Figure 5.10 Damped SDOF system with a rotating unbalance.

angle, θ, away from the true horizontal. As a result, we can represent θ as

$$\theta = \omega t \tag{5.47}$$

We need to work out a kinematic expression for the acceleration of the center of mass of the rotational component before writing the governing equations. We start with the acceleration of point O which can be written as

$$\vec{a}_O = \ddot{x}\vec{j} \tag{5.48}$$

where we are using the unit vectors $(\vec{i}, \vec{j})$ shown in the figure[2].

The remainder of the kinematic analysis proceeds by writing an expression for the position of m_r with respect to O.

$$\vec{P}_{m_r/O} = e \cos \theta \vec{i} + e \sin \theta \vec{j} \tag{5.49}$$

then differentiating to find

$$\vec{v}_{m_r/O} = -e\dot{\theta} \sin \theta \vec{i} + e\dot{\theta} \cos \theta \vec{j} \tag{5.50}$$

and differentiating again to get the relative acceleration

$$\vec{a}_{m_r/O} = (-e\ddot{\theta} \sin \theta - e\dot{\theta}^2 \cos \theta)\vec{i} + (e\ddot{\theta} \cos \theta - e\dot{\theta}^2 \sin \theta)\vec{j} \tag{5.51}$$

Recognizing that $\dot{\theta} = \omega$ and then $\ddot{\theta} = 0$ since ω is a constant and using the fact that $\theta = \omega t$ from Equation 5.47 lets us write the absolute acceleration of m_r as

$$\vec{a}_{m_r} = \vec{a}_O + \vec{a}_{m_r/O} = (-e\omega^2 \cos \omega t)\vec{i} + (\ddot{x} - e\omega^2 \sin \omega t)\vec{j} \tag{5.52}$$

We now consider the free body diagrams shown in Figure 5.11. For the rotating mass, m_r, we can write

$$\sum F_i : -F_H = m_r a_{m_{ri}} = -m_r e\omega^2 \cos \omega t \tag{5.53}$$

2 This is a good time to reinforce the implicit constraints that we use in modeling systems. For the machine to have only a vertical acceleration, there would need to be supports along its sides that don't let it move horizontally. Equally, there should be supports that stop it from moving out of the plane we are considering or from rotating. All of these supports exist in the real machine but we choose not to show them on the diagrams. We simply assume that the motion is constrained to the single, vertical, degree of freedom.

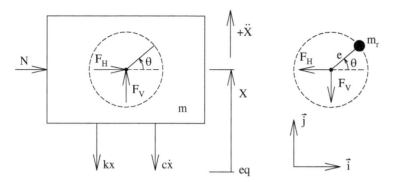

Figure 5.11 Free body diagram – Damped SDOF system with a rotating unbalance.

and

$$\sum F_j : -F_V = m_r a_{m_{rj}} = m_r(\ddot{x} - e\omega^2 \sin \omega t) \tag{5.54}$$

and, for the machine,

$$\sum F_i : N + F_H = (m - m_r)a_{Oi} = 0 \tag{5.55}$$

and

$$\sum F_j : -kx - c\dot{x} + F_V = (m - m_r)a_{Oj} = (m - m_r)\ddot{x} \tag{5.56}$$

Equations 5.53 and 5.55 are useful for showing that whatever is supporting the machine horizontally must be able to absorb the force

$$N = -m_r e\omega^2 \cos \omega t \tag{5.57}$$

thus showing that, even though we are ignoring the horizontal degree of freedom, it also needs to be considered in the design of machine supports.

Equations 5.54 and 5.56 can be combined to eliminate the force, F_V, so that we are left with the equation of motion

$$m\ddot{x} + c\dot{x} + kx = m_r e\omega^2 \sin \omega t \tag{5.58}$$

Equation 5.58 is similar to the previous forced vibration equations of motion we considered (Equations 5.13 and 5.25) but differs in having the magnitude of the applied force varying with the square of the forcing frequency. We will see that this makes a profound difference in the response.

We first write the equation of motion in standard form by dividing by the mass, m,

$$\ddot{x} + 2\zeta\omega_n\dot{x} + \omega_n^2 x = \frac{m_r e}{m}\omega^2 \sin \omega t \tag{5.59}$$

Now we consider the steady-state solution to Equation 5.59. Previously we were able to assume complex exponential forcing functions to make the analysis shorter. Here we are faced with a forcing function that is actually $\sin \omega t$ and not a complex exponential. We will therefore assume a steady-state solution that has both $\cos \omega t$ and $\sin \omega t$ in it and proceed to find an expression for the magnitude of the response[3].

---

3 It is easier to use the complex exponential to get the result, as we did earlier, but it is good practice to see this done both ways.

Let $x(t)$ be

$$x(t) = X_1 \cos \omega t + X_2 \sin \omega t \tag{5.60}$$

then

$$\dot{x}(t) = -\omega X_1 \sin \omega t + \omega X_2 \cos \omega t \tag{5.61}$$

and

$$\ddot{x}(t) = -\omega^2 X_1 \cos \omega t - \omega^2 X_2 \sin \omega t \tag{5.62}$$

Substituting into Equation 5.59 gives

$$(-\omega^2 X_1 \cos \omega t - \omega^2 X_2 \sin \omega t)$$
$$+ 2\zeta \omega_n (-\omega X_1 \sin \omega t + \omega X_2 \cos \omega t)$$
$$+ \omega_n^2 (X_1 \cos \omega t + X_2 \sin \omega t) = \frac{m_r e}{m} \omega^2 \sin \omega t \tag{5.63}$$

Since $\cos \omega t$ and $\sin \omega t$ are orthogonal functions, we can simply equate the $\cos \omega t$ terms on the left and right-hand sides of Equation 5.63, resulting in

$$(\omega_n^2 - \omega^2) X_1 + (2\zeta \omega \omega_n) X_2 = 0 \tag{5.64}$$

and then do the same thing for the $\sin \omega t$ terms, giving

$$-(2\zeta \omega \omega_n) X_1 + (\omega_n^2 - \omega^2) X_2 = \frac{m_r e}{m} \omega^2 \tag{5.65}$$

Equations 5.64 and 5.65 give two, simultaneous, algebraic, equations that can be used to solve for the magnitudes, X_1 and X_2. In fact, we don't really want to find them individually. We want the total magnitude of the response, X, where

$$X = \sqrt{X_1^2 + X_2^2} \tag{5.66}$$

and it turns out that we can find X fairly easily.

As a first step, we rewrite Equations 5.64 and 5.65 by dividing throughout by ω_n^2 to get

$$[1 - (\omega/\omega_n)^2] X_1 + [2\zeta (\omega/\omega_n)] X_2 = 0 \tag{5.67}$$

and then do the same thing for the $\sin \omega t$ terms, giving

$$-[2\zeta (\omega/\omega_n)] X_1 + [1 - (\omega/\omega_n)^2] X_2 = \frac{m_r e}{m} (\omega/\omega_n)^2 \tag{5.68}$$

then we square Equation 5.67

$$[1 - (\omega/\omega_n)^2]^2 X_1^2 + 4\zeta (\omega/\omega_n)[1 - (\omega/\omega_n)^2] X_1 X_2$$
$$+ [2\zeta (\omega/\omega_n)]^2 X_2^2 = 0 \tag{5.69}$$

and add it to the square of Equation 5.68

$$[2\zeta (\omega/\omega_n)]^2 X_1^2 - 4\zeta (\omega/\omega_n)[1 - (\omega/\omega_n)^2] X_1 X_2$$
$$+ [1 - (\omega/\omega_n)^2]^2 X_2^2 = \left(\frac{m_r e}{m}\right)^2 (\omega/\omega_n)^4 \tag{5.70}$$

to get

$$\{[1 - (\omega/\omega_n)^2]^2 + [2\zeta (\omega/\omega_n)]^2\}(X_1^2 + X_2^2) = \left(\frac{m_r e}{m}\right)^2 (\omega/\omega_n)^4 \tag{5.71}$$

We can then write

$$X^2 = (X_1^2 + X_2^2) = \frac{\left(\frac{m_r e}{m}\right)^2 (\omega/\omega_n)^4}{\{[1 - (\omega/\omega_n)^2]^2 + [2\zeta(\omega/\omega_n)]^2\}}$$

(5.72)

from which the amplitude, X, is

$$X = \frac{\left(\frac{m_r e}{m}\right)(\omega/\omega_n)^2}{\sqrt{[1 - (\omega/\omega_n)^2]^2 + [2\zeta(\omega/\omega_n)]^2}}$$

(5.73)

Finally, we define the dimensionless Amplitude Ratio to be

$$\frac{mX}{m_r e} = \frac{(\omega/\omega_n)^2}{\sqrt{[1 - (\omega/\omega_n)^2]^2 + [2\zeta(\omega/\omega_n)]^2}}$$

(5.74)

Figure 5.12 shows how the Amplitude Ratio varies with forcing frequency. The plot is similar to the other forced response plots in that it peaks near the undamped natural frequency for low damping ratios[4]. The response to a rotating unbalance has a forcing function whose magnitude grows with the forcing frequency squared (notice the $(\omega/\omega_n)^2$ term in the numerator) and the response, which for other systems declined at higher forcing frequencies, settles out at $mX/m_r e = 1$ for high forcing frequencies. Thus, once the machine is operating at a speed significantly above its undamped natural frequency, the amplitude of vibration will be the same for all higher operating speeds. The only concern then is the force transmitted to the supporting structure through the damper because of the high velocities that come with high frequencies.

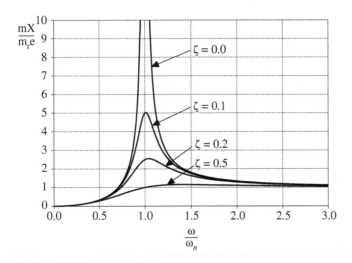

Figure 5.12 Amplitude Ratio – SDOF rotating unbalance.

4 The actual peak is at

$$(\omega/\omega_n) = \frac{1}{\sqrt{1 - 2\zeta^2}}$$

This is the only case where the peak response is at a forcing frequency above the undamped natural frequency.

5.5 Accelerometers

The measurement of acceleration is very important to the experimental characterization of vibrations. One type of transducer that is commonly used to measure acceleration is the *piezoelectric accelerometer*. A schematic of a typical piezoelectric accelerometer is shown in Figure 5.13. The basis for the design is the property of piezoelectric materials whereby they produce an electric potential in response to mechanical strain. If a piezoelectric material is bonded to two electrodes and then deformed, there will be a measurable voltage between the electrodes.

To measure acceleration, the accelerometer base is securely fixed to a vibrating structure so that the base experiences harmonic motion at the frequency of oscillation of the structure. This fits nicely into the harmonic base motion that we considered in Section 5.3. The mass is included in order to require that the base acceleration induce a force to pass through the piezoelectric material thereby causing strain and output voltage. The transmitted force will be equal to the mass multiplied by its absolute acceleration. The spring is there primarily to ensure that the piezoelectric material stays in compression at all times since these are typically brittle materials that can't withstand tensile loading. As a result, there is a significant preload in the spring when the system is in equilibrium and what we actually measure is the fluctuation in the material's output voltage from the equilibrium value. There will inevitably be damping in the system as well.

The model we use is shown in Figure 5.14. There is a subtle difference between this model and that used in Section 5.3. If you refer to Figure 5.6, you will see that we defined x to be the motion of the mass relative to its fixed equilibrium position so that its absolute acceleration was simply $\ddot{x}$. In Figure 5.14, x is defined to be the motion of the mass *relative to the base*. This means that the absolute acceleration of the mass must include the absolute acceleration of the base, $\ddot{y}$, plus the acceleration of the mass relative to the base, $\ddot{x}$. The reason for this change is that we want to see how the strain in the piezoelectric material is affected by the base acceleration and the relative motion between the mass and the base is a direct measure of the strain. We can therefore say that the output voltage will be proportional to the magnitude of $x(t)$.

The absolute acceleration of the mass is

$$a = (\ddot{y} + \ddot{x}) \tag{5.75}$$

as shown on the free body diagram in Figure 5.15.

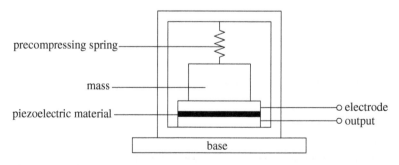

Figure 5.13 Schematic layout of a typical piezoelectric accelerometer.

Figure 5.14   Model of a typical piezoelectric accelerometer.

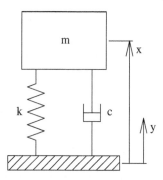

Figure 5.15   Free body diagram of the accelerometer model.

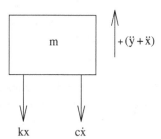

A vertical force balance yields

$$-c\dot{x} - kx = m(\ddot{y} + \ddot{x}) \tag{5.76}$$

giving the equation of motion as

$$m\ddot{x} + c\dot{x} + kx = -m\ddot{y} \tag{5.77}$$

This can be written in standard form as

$$\ddot{x} + 2\zeta\omega_n\dot{x} + \omega_n^2 x = -\ddot{y} \tag{5.78}$$

Let $y(t)$ be harmonic with amplitude, Y, and frequency, ω, so we can write

$$y(t) = Ye^{i\omega t} \tag{5.79}$$

and then differentiate twice to get

$$\ddot{y}(t) = -\omega^2 Ye^{i\omega t} \tag{5.80}$$

We then assume, in the usual fashion, that the steady state response will be

$$x(t) = Xe^{i\omega t} \tag{5.81}$$

Equation 5.81 is then differentiated twice giving, after substitution of the results into Equation 5.78,

$$(-\omega^2 + 2\zeta\omega_n\omega i + \omega_n^2)Xe^{i\omega t} = \omega^2 Ye^{i\omega t} \tag{5.82}$$

We can then solve for the relationship between X, representing strain, and $\omega^2 Y$, representing the acceleration we are trying to measure. Defining $A = \omega^2 Y$ to be the magnitude of the acceleration, we can write

$$X = \frac{A}{(\omega_n^2 - \omega^2) + (2\zeta\omega_n\omega i)} \tag{5.83}$$

Factoring out ω_n^2 results in

$$X = \frac{A}{\omega_n^2\{[1 - (\omega/\omega_n)^2] + (2\zeta\omega/\omega_n i)\}} \tag{5.84}$$

and, taking the magnitudes, we see how the strain or output voltage is related to the acceleration amplitude

$$X = \frac{A}{\omega_n^2 \sqrt{[1 - (\omega/\omega_n)^2]^2 + (2\zeta\omega/\omega_n)^2}} \tag{5.85}$$

Ideally, we would want the output of a transducer to be proportional to the signal that is being measured. That is

$$X = CA \tag{5.86}$$

whereas Equation 5.85 indicates that this seems not to be the case. If, however, we consider the case where ω/ω_n is much smaller than one, the square root term in the denominator of Equation 5.85 is approximately equal to one and we get

$$X \approx \frac{1}{\omega_n^2}A \tag{5.87}$$

where X is proportional to A and the constant of proportionality is $C = 1/\omega_n^2$.

Figure 5.16 shows the percentage error in the assumed proportionality as a function of frequency ratio for different values of damping ratios. The error is acceptable so long as the measurement is taken at less than 10% of the undamped natural frequency and even up to 20% of ω_n if the damping ratio is 0.6 or 0.7. In fact, the limited frequency range is not a particularly onerous restriction because the accelerometers have relatively small masses and

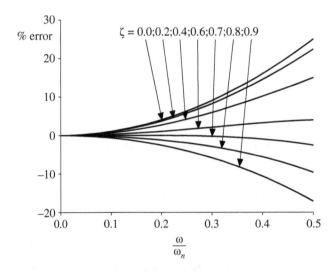

Figure 5.16 Piezoelectric accelerometer – % error.

stiff springs so their natural frequencies are usually in the 10s of kilohertz range so measuring typical structural vibrations that are less than 100 Hz is not a problem. Accelerometer manufacturers will specify the natural frequencies of their instruments.

The constant of proportionality, C, in Equation 5.86 is measured by mounting the accelerometer on an accelerometer calibrator that produces 1g at a known frequency and recording that measurement as a reference.

Exercises

5.1 A simple device for measuring the magnitudes of earthquakes can be constructed by using springs to mount a mass within a rigid box as shown in the figure. A needle is fixed to the mass and points to a scale on the side of the box. As the box moves up and down with the ground ($y(t)$ on the figure), the motion of the internal mass relative to the box ($x(t)$ on the figure) is indicated by the reading on the scale. Assume that The earthquake causes a harmonic ground motion, $y(t) = Y \sin \omega t$, and that the box follows this perfectly.

(a) Derive an expression for the steady-state amplitude that would be read on the scale (i.e. the magnitude of $x(t)$). Express it as a function of the ground amplitude, Y, the forcing frequency, ω, the mass, m, and the stiffness, k.

(b) Sketch a graph of the amplitude ratio ($|X/Y|$) as a function of the frequency ratio (ω/ω_n). Recognizing that the instrument is only usable when the amplitude ratio is constant with respect to frequency, indicate a range of frequency ratios where the measured value of X can be used to predict Y with an error of less than one percent.

(c) A typical earthquake provides a forcing frequency near 3 Hz. Given that m = 20 kg, what spring stiffness, k, should be used?

Figure E5.1

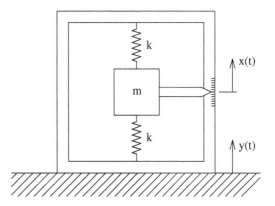

5.2 A 225-kg electric motor is supported by an elastic pad that deflects 6 mm under the weight of the motor and provides no damping. When the motor operates at 1750 RPM, it oscillates with an amplitude of 2.5 mm. In an attempt to reduce the amplitude of the vibration, the operators plan to place a 680-kg steel platform between the motor and the pad. What would be the expected amplitude of oscillation at 1750 RPM with the platform in place?

5.3 The figure shows a cart of mass, m, that is attached to the ground at point, A, by a spring of stiffness, k, and a damper with damping coefficient, c. The motion of the cart is forced by a harmonic motion, $y(t)$, at the end of another damper with coefficient, c. $y(t)$ has amplitude, Y, and frequency, ω.

(a) Write the equation of motion for the system using $x(t)$ as the degree of freedom.

(b) Derive an expression for the steady-state amplitude of $x(t)$.

(c) Derive an expression for the magnitude of the force transmitted to the wall at point A.

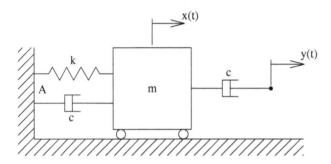

Figure E5.3

5.4 A printed circuit board with a mass of 0.5 kg is mounted to a base using an isolator with no damping. During shipping, the base is subjected to a harmonic motion of amplitude 2 mm and frequency 2 Hz. Choose an isolator stiffness so that the displacement of the printed circuit board will be no more than 5% of the base motion.

5.5 A piece of vibration sensitive equipment is mounted on four rubber pads connected to a concrete floor. The mass of the machine is 1000 kg and the vertical stiffness of a single rubber pad is 4×10^6 N/m.

(a) What is the undamped natural frequency of vertical vibrations of the machine?

(b) The concrete floor vibrates at 20 Hz due to a compressor running in a neighboring room. What is the ratio of the amplitude of vibration of the machine to the amplitude of vibration of the floor?

(c) The rubber pads have some inherent damping which can be approximated as viscous. What c value is required per pad to reduce the amplitude of vibration of the machine to one half that experienced in part (b)?

5.6 A machine with total mass, m, has a rotating unbalance. The unbalanced mass is m_r and the eccentricity is e. The system is supported by elastic elements having a total stiffness, k, and negligible damping. The rotational speed is ten to fifteen times greater than the natural frequency.

Derive an expression for the magnitude of the dynamic force transmitted to the floor. How much does the force vary in the range of rotational speeds specified.

5.7 A variable-speed electric motor with a mass of 200 kg has a rotating unbalance due to manufacturing errors. The motor is mounted on rubber mounts that have an effective stiffness of 10 kN/m and an effective damping ratio of 0.15.

(a) Show that the ratio of the force transmitted to the foundation, F_T, to the unbalanced exciting force, $m_r e\omega^2$, is

$$\frac{F_T}{m_r e\omega^2} = \sqrt{\frac{1 + \left(2\zeta\frac{\omega}{\omega_n}\right)^2}{\left[1 - \left(\frac{\omega}{\omega_n}\right)^2\right]^2 + \left(2\zeta\frac{\omega}{\omega_n}\right)^2}}$$

(b) Find the speed range, in RPM, over which the amplitude of the force transmitted to the foundation will be larger than the exciting force.

(c) Over what speed range will the transmitted force amplitude be less than 10% of the exciting force amplitude?

5.8 A vibrometer (a machine for measuring vibrations) has been designed to measure the vertical motions of a flat surface. The design is relatively simple, consisting of a rigid arm of negligible mass pinned to a frame at one end and supporting a point mass, m, at the end with the pointer. As the frame responds to the harmonic surface motions, $y(t) = Y \sin \omega t$, the arm rotates and the vertical motion of the pointer provides a scaled measurement of the amplitude of motion of the surface.

(a) Derive the linearized equation of motion of the for the vibrometer using the angular motion θ as the single degree of freedom.

(b) If the surface moves at the frequency $\omega = \sqrt{k/m}$, what fraction of Y will the pointer indicate?

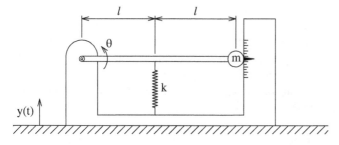

Figure E5.8

5.9 A simple pendulum (mass $= m$, length $= \ell$) is supported by a small wheel rolling on a horizontal surface (point A). The wheel has negligible mass and is forced to oscillate horizontally with a known amplitude, Y, and frequency, ω, so that its motion can be described by

$$y(t) = Y \sin \omega t$$

(a) Show that the nonlinear equation of motion for the single degree of freedom, θ, can be written as

$$m\ell^2\ddot{\theta} + mg\ell \sin \theta = -m\ddot{y}\ell \cos \theta$$

(b) Linearize the equation of motion about the equilibrium where $\theta = 0$ and write an expression for the undamped natural frequency.

(c) If the pendulum has length, $\ell = 100\,\text{cm}$, the disturbance has amplitude, $Y = 1\,\text{cm}$, and the forcing frequency is

$$\omega = \sqrt{\frac{3g}{\ell}}$$

What will the amplitude of θ be?

Figure E5.9

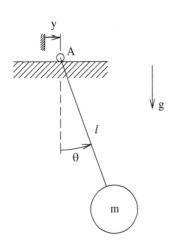

5.10 A 100 kg machine is mounted on an elastic foundation and an experiment is run to determine the stiffness and damping properties of the foundation. When the machine is excited with a harmonic force of 100 N at several frequencies, the maximum steady-state amplitude is observed to be 0.5 mm and it occurs at 100 rad/s.

(a) Assume small damping to get approximations for the stiffness (k in N/m) and damping (c in N·s/m) that the foundation is providing.

(b) Given the damping ratio that you found, comment on whether or not the approximations used are reasonable.

5.11 Consider the case of a machine with a rotating unbalance mounted to the floor via a spring and damper. Show that the ratio of the magnitude of the force transmitted through the damper to the magnitude of that transmitted through the spring is

$$\frac{|F_c|}{|F_k|} = 2\zeta \left(\frac{\omega}{\omega_n}\right)$$

6

Damping

Damping involves the removal of energy from vibrating systems. There are many mechanisms that provide damping and no system is immune from them. Friction, air resistance, and the propensity for deformed materials to generate internal heat all take energy out of the system and provide damping.

6.1 Linear Viscous Damping

The classic model of a damping element is one in which the force generated is proportional to the rate of change of length of the element. The rate of change of length is the relative velocity across the damper. "Proportional" implies linearity and a force proportional to speed implies laminar, viscous flow. As a result, these elements are often referred to as "linear viscous dampers" or simply "viscous dampers". These elements were previously defined in Section 1.2.4 where Rayleigh's dissipation function was introduced and some of that material is repeated here.

Figure 6.1 shows a system where a body is attached to ground by a damper. The body is moving to the right with speed v and the damping coefficient (constant of proportionality) is c. The physical connection of the damper to both the ground and the body dictates that the rate of change of length of the damper is equal to the speed v. The force in the damper will therefore be $F_d = cv$. The direction of the force will be such that it causes the damper to increase in length as shown in the lower part of Figure 6.1. By Newton's 3rd Law, the force on the body must be equal and opposite to the force acting on the damper. The force F_d therefore acts to the left on the body. In other words, the damping force opposes the velocity of the body.

We will be categorizing damping by the amount of energy removed per cycle of motion. We can calculate that value analytically for the linear viscous damper. To do this, we visualize an experiment (Figure 6.2) where we attach one end of a damper to ground and apply a force to the free end that makes that end experience motion according to

$$x(t) = X \sin \omega t \tag{6.1}$$

where the amplitude, X, and the frequency, ω, are both known. The relative velocity across the damper will then be

$$v = \dot{x}(t) = \omega X \cos \omega t \tag{6.2}$$

Introduction to Mechanical Vibrations, First Edition. Ronald J. Anderson.
© 2020 John Wiley & Sons Ltd. Published 2020 by John Wiley & Sons Ltd.
Companion website: www.wiley.com/go/anderson/introduction-to-vibrations

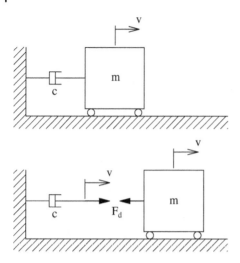

Figure 6.1   A linear viscous damper.

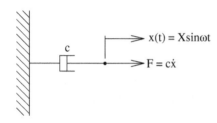

Figure 6.2   Test setup for energy removed by a linear viscous damper.

and the force required to generate the prescribed displacement is

$$F = cv = c\,\omega X \cos \omega t \tag{6.3}$$

Figure 6.3 shows the result of plotting force versus displacement for the viscous damper test. Each cycle of the test produces a complete ellipse as shown in the figure. The range of displacement is $-X \le x \le +X$ and the force varies over $-c\omega X \le F \le +c\omega X$. The ellipse is called a *hysteresis loop* and the area inside the loop is the integral of force multiplied by distance over a cycle of motion. The area is therefore the work done on the damper per cycle or, equivalently, the *energy removed from the system per cycle of motion* by the damper.

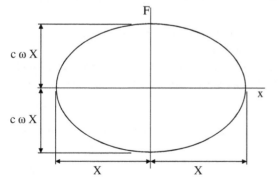

Figure 6.3   Hysteresis loop for a linear viscous damper.

One cycle of motion corresponds to ωt going from 0 to 2π. The work done, W, in forcing the damper through that cycle can be calculated by the integral of Fdx over the cycle. We start with

$$W = \int_{\omega t=0}^{\omega t=2\pi} Fdx \tag{6.4}$$

and then substitute

$$F = cv = c\left(\frac{dx}{dt}\right) \tag{6.5}$$

to get

$$W = \int_{\omega t=0}^{\omega t=2\pi} c\left(\frac{dx}{dt}\right)dx \tag{6.6}$$

We can multiply and divide by dt in Equation 6.6, yielding

$$W = \int_{\omega t=0}^{\omega t=2\pi} c\left(\frac{dx}{dt}\right)dx\left(\frac{dt}{dt}\right) \tag{6.7}$$

and then rearrange to get

$$W = \int_{\omega t=0}^{\omega t=2\pi} c\left(\frac{dx}{dt}\right)\left(\frac{dx}{dt}\right)dt = \int_{\omega t=0}^{\omega t=2\pi} c\left(\frac{dx}{dt}\right)^2 dt \tag{6.8}$$

We have an expression for the velocity, dx/dt, from Equation 6.2 that can be substituted here to give

$$W = \int_{\omega t=0}^{\omega t=2\pi} c(\omega X \cos \omega t)^2 dt \tag{6.9}$$

where we can factor out some constant values and change dt to $\frac{d(\omega t)}{\omega}$ in order to agree with the limits on the integration, giving

$$W = c\omega^2 X^2 \int_{\omega t=0}^{\omega t=2\pi} \cos^2 \omega t \frac{d(\omega t)}{\omega} \tag{6.10}$$

or

$$W = c\omega X^2 \int_{\omega t=0}^{\omega t=2\pi} \cos^2 \omega t \, d(\omega t) \tag{6.11}$$

The integral in Equation 6.11 is of the form

$$\int_0^{2\pi} \cos^2 u \, du \tag{6.12}$$

You can either look this integral up in a table of integrals or use integration by parts to evaluate it. Integration by parts starts with the differential of a product

$$d(\alpha\beta) = \alpha d\beta + \beta d\alpha$$

then rearranges

$$\alpha d\beta = d(\alpha\beta) - \beta d\alpha$$

integrates

$$\int \alpha d\beta = \int d(\alpha\beta) - \int \beta d\alpha$$

and evaluates one of the integrals to leave the following formula

$$\int \alpha d\beta = \alpha\beta - \int \beta d\alpha$$

For the integral being considered, we let

$$\alpha = \cos u \Rightarrow d\alpha = -\sin u \, du$$

and

$$d\beta = \cos u \, du \Rightarrow \beta = \sin u$$

then, using the integration by parts formula, we write

$$\int (\cos u)(\cos u \, du) = \cos u \sin u - \int (\sin u)(-\sin u \, du)$$

or

$$\int \cos^2 u \, du = \cos u \sin u + \int \sin^2 u \, du$$

Then use the identity $\sin^2 u + \cos^2 u = 1$ or $\sin^2 u = 1 - \cos^2 u$ to rewrite it as

$$\int \cos^2 u \, du = \cos u \sin u + \int (1 - \cos^2 u) \, du$$

from which

$$2 \int \cos^2 u \, du = \cos u \sin u + \int du$$

or

$$\int \cos^2 u \, du = \frac{1}{2}[\cos u \sin u + u]$$

The result is

$$\int_0^{2\pi} \cos^2 u \, du = \frac{1}{2}[\cos u \sin u + u]_0^{2\pi} = \pi \tag{6.13}$$

Giving, at last, the expression for the energy dissipated per cycle by a viscous damper as

$$W = \pi c \omega X^2 \tag{6.14}$$

Figure 6.4 shows experimental hysteresis loops for two different shock absorbers. One is designed for an intercity bus and the other for a midsized car. The shock absorbers were cycled several times to warm up the oil and the tests shown here consist of four loops for each of the shock absorbers. The repeatability is excellent after the warm up so that the loops are nearly perfectly overlapping.

Table 6.1 presents the effective viscous damping coefficients derived from the loops. The procedure is to perform a numerical integration to find the area inside a loop (being careful to use consistent units – displacement in m, frequency in rad/s) and then equating the area to $\pi c_{eq} \omega X^2$ and finding c_{eq}. It is important when doing tests like this to use amplitudes and frequencies that are close to those that will be experienced during normal operation of the equipment. The values of c_{eq} are very dependent on frequency and amplitude for some damping elements.

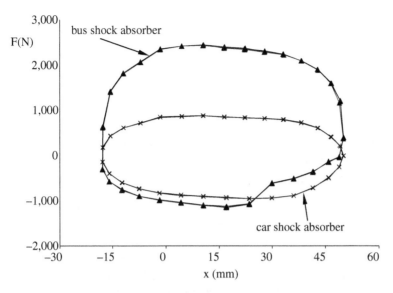

Figure 6.4 Experimental hysteresis loops for two shock absorbers.

Table 6.1 Effective viscous damping coefficients.

vehicle	Frequency Hz	Amplitude mm	c_{eq} kN · s/m
bus	5.1	± 34.1	1.66
car	5.1	± 34.1	0.86

6.2 Coulomb or Dry Friction Damping

Friction exists everywhere and is a major source of damping in systems. Frictional forces only exist between bodies that are in contact with each other and the magnitude of the friction force depends on surface properties and the force pushing the bodies together. We characterize the surface properties using *coefficients of friction* (μ_s = static coefficient of friction and μ_k = kinetic[1] coefficient of friction with $\mu_s > \mu_k$). The force acting between the bodies that is perpendicular to their plane of contact is the *normal force* and we denote that force as N.

The direction of the friction force is an important consideration. Figure 6.5 shows the force of friction acting on the body[2] for the three possible cases. On the left and right you will see that, if the body has a velocity relative to the ground, there is a friction force equal to μN and it will be acting in a direction that attempts to reduce the velocity. It will therefore be aligned with and opposed to the velocity vector.

---

1 It is common practice to refer to the kinetic coefficient of friction as simply μ since it is used so much more often than the static coefficient of friction. That will be done from here onward.
2 By Newton's Third Law, the force of friction acting on the ground will be equal and opposite.

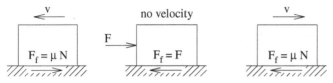

Figure 6.5 Friction force magnitude and direction.

The center figure answers the question – *What is the friction force if there is no relative velocity?* The figure shows a force, F, being applied to the body while it is stationary. It also shows the friction force as being equal and opposite to the applied force. In fact, the friction force takes on any value that is necessary to stop the body from moving until it needs to be greater than $\mu_s N$ to do that. Then the body begins to slide and the friction force reverts to being held constant at μN and opposing the velocity.

Consider now a mass that is free to slide on a surface with friction and that has a spring attaching it to ground. Figure 6.6 shows the system in motion. The situation shown is one where equilibrium has the spring at its undeflected length with no force acting on it. It is then pulled in the positive x-direction until it reaches its initial displacement, x_0, and released from rest. The force that the spring exerts on the body when the system is released will be kx_0 to the left. If $kx_0 \leq \mu_s N$, the spring cannot overcome the friction force and the body will remain stationary at the position $x = x_0$. If, however, $kx_0 > \mu_s N$, the spring force will overcome the friction force and the body will start to slide toward its equilibrium position. As a result, the velocity will be to the left and the friction force (μN) will oppose it.

The equation of motion for the system is

$$m\ddot{x} + kx = \mu N \qquad (6.15)$$

This is a non-homogeneous, linear, ordinary, differential equation with a constant on the right-hand side and we can solve it.

The complementary solution is

$$x_c(t) = A\cos\omega_n t + B\sin\omega_n t \qquad (6.16)$$

and the particular solution is

$$x_p(t) = X = \text{a constant to match the constant right-hand side} \qquad (6.17)$$

Clearly we must have

$$X = \frac{\mu N}{k} \qquad (6.18)$$

Figure 6.6 A system with Coulomb friction.

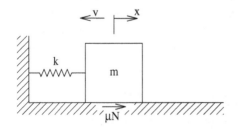

so that the solution is $x(t) = x_c(t) + x_p(t)$ or

$$x(t) = A \cos \omega_n t + B \sin \omega_n t + \frac{\mu N}{k} \tag{6.19}$$

Applying the initial conditions, $x(0) = x_0$ and $\dot{x}(0) = 0$, gives

$$A = x_0 - \frac{\mu N}{k} \tag{6.20}$$

and

$$B = 0 \tag{6.21}$$

so that the solution becomes

$$x(t) = \left(x_0 - \frac{\mu N}{k} \right) \cos \omega_n t + \frac{\mu N}{k} \tag{6.22}$$

It is important to realize that the equation of motion we are working with is only valid when the velocity of the body is to the left and that will be true for only a half cycle of motion. The body will continue moving to the left until $\omega_n t = \pi$ and then it will stop with

$$x(t) = -x_0 + 2\frac{\mu N}{k} \tag{6.23}$$

We need to check again whether the spring force can overcome the static friction force with this displacement. If it can, the body will begin to move to the right and the friction force will reverse direction so that the equation of motion becomes

$$m\ddot{x} + kx = -\mu N \tag{6.24}$$

This equation is valid for another half cycle of motion and we will again lose $2\mu N/k$ from the peak amplitude.

Figure 6.7 shows the response. The system will continue moving from peak to peak until the spring force can't overcome the static friction force ($x(t) = x_f$ in the figure where $kx_f \leq \mu_s N$) and all motion will stop. Until this happens, the body will oscillate at its undamped natural frequency, ω_n, and will lose amplitude at the constant rate of $4\mu N/k$ per cycle.

The straight-line decay seen in Figure 6.7 is characteristic of Coulomb or dry friction damping and is often seen in practice since there is friction everywhere. This is the only type of damping that doesn't affect the natural frequency of the system and that causes

Figure 6.7 Response with Coulomb friction.

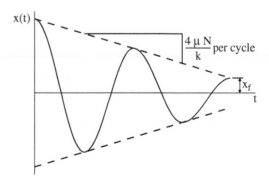

the motion to stop completely. Contrast this with the linear viscous damping that we use nearly all of the time in our analyses to predict an exponential decay that asymptotically approaches zero but never gets there.

6.3 Logarithmic Decrement

We considered the response of underdamped systems to initial conditions in Subsection 3.3.3 where we showed in Equation 3.72 that

$$x(t) = e^{-\zeta\omega_n t}\left[x_0 \cos\omega_d t + \frac{v_0 + \zeta\omega_n x_0}{\omega_d}\sin\omega_d t\right] \tag{6.25}$$

which is plotted in Figure 6.8

We could equally as well have written Equation 6.25 as a function with an amplitude, X, and a phase angle, ϕ, so that

$$x(t) = Xe^{-\zeta\omega_n t}\cos(\omega_d t + \phi)) \tag{6.26}$$

and we will use this form for the discussion of logarithmic decrement.

Consider the peak labeled x_i in Figure 6.8. Since this is a peak value, it must occur at a time, t_i, where $\cos(\omega_d t_i + \phi)) = 1$. This may not seem immediately obvious, but look at Equation 6.26 keeping in mind that the cosine function varies between plus and minus one and we are looking at a positive peak value. As a result, we can say that

$$x_i = Xe^{-\zeta\omega_n t_i} \tag{6.27}$$

We now proceed to the next peak and make the same argument at time, t_j, and write

$$x_j = Xe^{-\zeta\omega_n t_j} \tag{6.28}$$

Now take the ratio of Equation 6.27 to 6.28 and write

$$\frac{x_i}{x_j} = \frac{Xe^{-\zeta\omega_n t_i}}{Xe^{-\zeta\omega_n t_j}} = e^{\zeta\omega_n (t_j - t_i)} \tag{6.29}$$

Now note that $(t_j - t_i)$ is equal to the period, T, and that

$$T = \frac{2\pi}{\omega_d} \tag{6.30}$$

Figure 6.8 The underdamped response.

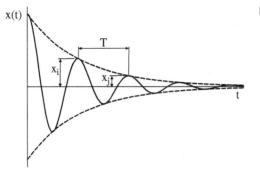

allowing Equation 6.29 to be written as

$$\frac{x_i}{x_j} = e^{\zeta \omega_n (2\pi/\omega_d)} \tag{6.31}$$

We now take the natural logarithm of both sides of Equation 6.31 to get

$$\ln\left(\frac{x_i}{x_j}\right) = \frac{2\pi\zeta\omega_n}{\omega_d} \tag{6.32}$$

We know that the damped natural frequency is related to the undamped natural frequency by

$$\omega_d = \omega_n\sqrt{1 - \zeta^2} \tag{6.33}$$

and we define the *logarithmic decrement*, δ, to be

$$\delta = \ln\left(\frac{x_i}{x_j}\right) \tag{6.34}$$

so that Equation 6.32 can be written as

$$\delta = \frac{2\pi\zeta\omega_n}{\omega_n\sqrt{1-\zeta^2}} = \frac{2\pi\zeta}{\sqrt{1-\zeta^2}} \tag{6.35}$$

Equation 6.35 can be solved for the damping ratio, ζ, as follows. First square both sides

$$\delta^2 = \frac{4\pi^2\zeta^2}{1 - \zeta^2} \tag{6.36}$$

then multiply both sides by $1 - \zeta^2$

$$\delta^2(1 - \zeta^2) = 4\pi^2\zeta^2 \tag{6.37}$$

from which

$$\delta^2 = (4\pi^2 + \delta^2)\zeta^2 \tag{6.38}$$

giving

$$\zeta^2 = \frac{\delta^2}{(4\pi^2 + \delta^2)} \tag{6.39}$$

and, finally

$$\zeta = \frac{\delta}{\sqrt{4\pi^2 + \delta^2}} \tag{6.40}$$

The logarithmic decrement method is used to estimate damping ratios from experimental results. Given an experimental result resembling Figure 6.8, you can use any two neighboring peaks to generate a quick estimate of the level of damping in the system.

Exercises

6.1 What is the damping ratio of the signal shown in the figure? Explain how you know that the damping is not due to dry friction.

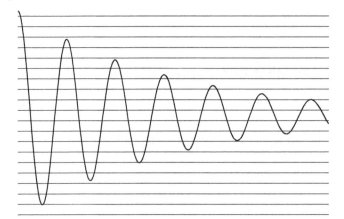

Figure E6.1

6.2 Show that the equivalent viscous damping coefficient for a block of mass, m, forced to undergo harmonic motion, $x(t) = X \sin \omega t$, while resting on a surface with coefficient of friction, μ, is

$$c_{eq} = \frac{4\mu mg}{\pi \omega X}$$

6.3 A new damping device has been tested and has a force versus displacement hysteresis loop as shown. The test was done at a frequency of 1 Hz. What is the equivalent viscous damping coefficient, c, that you would use to model this device?

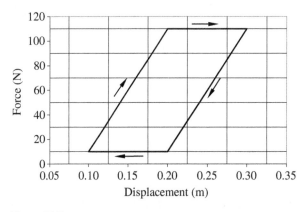

Figure E6.3

6.4 A machine with a rotating unbalance runs at 1800 RPM. The total mass of the machine is 1000 kg and it is mounted on four rubber pads connected to a concrete floor. The vertical stiffness of a single rubber pad is 1×10^7 N/m.
(a) What is the undamped natural frequency of vertical vibrations of the machine?
(b) The rotating mass is 200 kg and the eccentricity is 0.5 mm. Assuming no damping, what is the amplitude of vibration of the machine?

(c) The rubber pads have some inherent damping which can be approximated as viscous. What c value is required per pad to reduce the amplitude of vibration of the machine to one tenth of that experienced in part (b)?

(d) One of the rubber pads has been tested in the lab by applying a sinusoidal displacement with amplitude 0.1 mm at a frequency of 30 Hz. The tests indicate the work done per cycle is 0.38 Joules. Is this sufficient to provide the c value calculated in part (c) or will additional damping be required?

6.5 The resisting force due to fluid drag is proportional to the square of the velocity. This observation leads to the *velocity-squared* damping model, which gives a drag force of

$$F = \frac{1}{2} \rho \, C_d \, A \, sgn(v) \, v^2$$

where ρ is the fluid mass density, C_d is the drag coefficient, and A is the object's area normal to the flow.

Let $b = \rho \, C_d \, A/2$ and follow the procedure in Section 6.1 to find the equivalent viscous damping coefficient for velocity-squared damping.

Notes:

1 The fact that the drag force changes sign (thus the $sgn(v)$) as v reverses but the force is expressed as a function of v^2 which never changes sign, means that you have to integrate only over a quarter cycle ($0 \le \omega t \le \pi/2$) and then multiply the result by 4 to get the total work done per cycle.

2 The integral you need is

$$\int \cos^3 u \, du = \frac{1}{3} \sin u \, (\cos^2 u + 2)$$

7

Systems with More than One Degree of Freedom

A degree of freedom is a coordinate, either translational or rotational, required to describe the motion of a mass in a system. We could have, for example, the problem of describing the general motion of a rigid body constrained to move on a plane. A complete specification of its motion would require two orthogonal translational coordinates to locate its center of mass plus a single rotational coordinate to specify its orientation on the plane. There are therefore three degrees of freedom for the body. If we released the constraint that it must move on a plane, we would need to deal with a rigid body moving in three-dimensional space and would find ourselves with six degrees of freedom (three translations and three rotations).

Alternatively, we might be dealing with a system that has multiple bodies, each with a single degree of freedom so that we need as many degrees of freedom as there are bodies.

In any case, we will now be dealing with NDOF systems – *systems with N degrees of freedom.*

We start with a discussion of systems with two degrees of freedom. These systems are relatively easy to consider and have all of the properties of systems with any number of degrees of freedom.

7.1   2DOF Undamped Free Vibrations – Modeling

We start with a 2DOF system without damping as shown in Figure 7.1. There are two masses, m_1 and m_2, that are constrained to have one-dimensional, vertical, motion only. We can define the two degrees of freedom as x_1 and x_2, the vertical motions away from the equilibrium positions of the masses. The effects of gravity can be ignored because they are completely canceled out by the preloads in the springs when the system is in equilibrium. As with 1DOF systems, we are only concerned with things that change when we move away from equilibrium and gravitational forces and preloads in springs don't change.

Figure 7.2 shows free body diagrams for the system. The masses are assumed to have moved away from equilibrium in the positive sense and their positive acceleration directions are shown. The forces acting on the masses are the portions of the forces that change due to motion around equilibrium (i.e. the forces that act in addition to the preload forces). The free body diagrams are drawn as if all of the springs are in tension. That is, both springs are pulling on the masses rather than pushing on them.

Introduction to Mechanical Vibrations, First Edition. Ronald J. Anderson.
© 2020 John Wiley & Sons Ltd. Published 2020 by John Wiley & Sons Ltd.
Companion website: www.wiley.com/go/anderson/introduction-to-vibrations

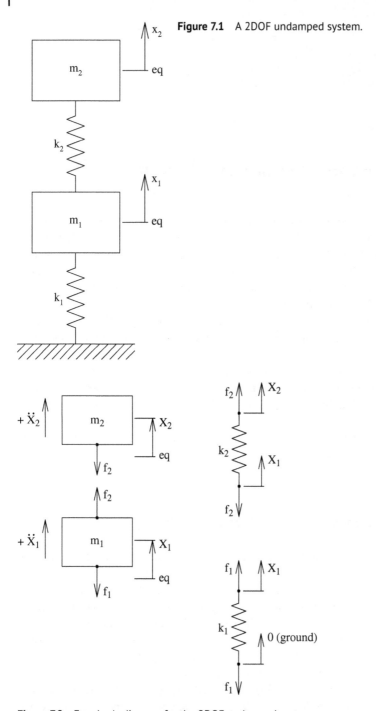

Figure 7.1 A 2DOF undamped system.

Figure 7.2 Free body diagram for the 2DOF undamped system.

The force balance equations on the two masses can be written as:

on m_1

$$+ \uparrow \sum F : f_2 - f_1 = m_1 \ddot{x}_1 \tag{7.1}$$

and, on m_2

$$+ \uparrow \sum F : -f_2 = m_2 \ddot{x}_2 \tag{7.2}$$

The free body diagrams of the springs show them in tension and the tensile deflection expressions (i.e. the deflections that make the springs longer) are

$$\delta_1 = x_1 - 0 = x_1 \tag{7.3}$$

and

$$\delta_2 = x_2 - x_1 \tag{7.4}$$

so that the spring forces are

$$f_1 = k_1 x_1 \tag{7.5}$$

and

$$f_2 = k_2 (x_2 - x_1) \tag{7.6}$$

Substituting Equations 7.5 and 7.6 into 7.1 and 7.2 gives

$$k_2 (x_2 - x_1) - k_1 x_1 = m_1 \ddot{x}_1 \tag{7.7}$$

and

$$-k_2 (x_2 - x_1) = m_2 \ddot{x}_2 \tag{7.8}$$

which can be rearranged to

$$m_1 \ddot{x}_1 + (k_1 + k_2) x_1 - k_2 x_2 = 0 \tag{7.9}$$

and

$$m_2 \ddot{x}_2 - k_2 x_1 + k_2 x_1 = 0 \tag{7.10}$$

Equations 7.9 and 7.10 can be written in matrix form as

$$\underbrace{\begin{bmatrix} m_1 & 0 \\ 0 & m_2 \end{bmatrix}}_{\text{mass matrix}} \begin{Bmatrix} \ddot{x}_1 \\ \ddot{x}_2 \end{Bmatrix} + \underbrace{\begin{bmatrix} (k_1 + k_2) & -k_2 \\ -k_2 & k_2 \end{bmatrix}}_{\text{stiffness matrix}} \begin{Bmatrix} x_1 \\ x_2 \end{Bmatrix} = \begin{Bmatrix} 0 \\ 0 \end{Bmatrix} \tag{7.11}$$

The general form of the equations of motion for undamped, free vibrations of a 2DOF system is then

$$[M]\{\ddot{x}\} + [K]\{x\} = \{0\} \tag{7.12}$$

where $[M]$ is the *mass matrix*, $[K]$ is the *stiffness matrix*, and $\{x\}$ is the vector of degrees of freedom. An undamped, free vibration, model of a system with N degrees of freedom will also have equations of motion of this form with the only difference being that the matrices are N×N rather than 2×2 as they are here.

7.2 2DOF Undamped Free Vibrations – Natural Frequencies

We now consider the possible forms of x_1 to x_N that can be used as solutions to sets of linear, ordinary, differential equations such as Equation 7.12. One thing is immediately obvious – every equation in the set adds together x_1 to x_N and the second derivatives, $\ddot{x}_1$ to $\ddot{x}_N$, multiplied by constants, and equates the sum to zero. This can only happen if x_i has the same form as x_j for all i and j and $\ddot{x}_i$ and $\ddot{x}_j$ match them exactly. If this doesn't happen, Equation 7.12 cannot have a solution.

We therefore are restricted to harmonic solutions where all of the x_i move with the same frequency but at different amplitudes. For the 2DOF system, we assume

$$x_1(t) = X_1 e^{i\omega t} \tag{7.13}$$

and

$$x_2(t) = X_2 e^{i\omega t} \tag{7.14}$$

Taking two derivatives, we find

$$\ddot{x}_1(t) = -\omega^2 X_1 e^{i\omega t} \tag{7.15}$$

and

$$\ddot{x}_2(t) = -\omega^2 X_2 e^{i\omega t} \tag{7.16}$$

Substituting into Equation 7.11 yields

$$\begin{bmatrix} -\omega^2 m_1 & 0 \\ 0 & -\omega^2 m_2 \end{bmatrix} \begin{Bmatrix} X_1 e^{i\omega t} \\ X_2 e^{i\omega t} \end{Bmatrix} + \begin{bmatrix} (k_1 + k_2) & -k_2 \\ -k_2 & k_2 \end{bmatrix} \begin{Bmatrix} X_1 e^{i\omega t} \\ X_2 e^{i\omega t} \end{Bmatrix} = \begin{Bmatrix} 0 \\ 0 \end{Bmatrix} \tag{7.17}$$

Or,

$$\begin{bmatrix} -\omega^2 m_1 + (k_1 + k_2) & -k_2 \\ -k_2 & -\omega^2 m_2 + k_2 \end{bmatrix} \begin{Bmatrix} X_1 \\ X_2 \end{Bmatrix} e^{i\omega t} = \begin{Bmatrix} 0 \\ 0 \end{Bmatrix} \tag{7.18}$$

Since $e^{i\omega t}$ is never zero, it can be canceled out, leaving

$$\begin{bmatrix} -\omega^2 m_1 + (k_1 + k_2) & -k_2 \\ -k_2 & -\omega^2 m_2 + k_2 \end{bmatrix} \begin{Bmatrix} X_1 \\ X_2 \end{Bmatrix} = \begin{Bmatrix} 0 \\ 0 \end{Bmatrix} \tag{7.19}$$

Equation 7.19 is a set of two linear, algebraic equations with three unknowns: ω, X_1, and X_2. That is, we don't know the frequency and we don't know the amplitudes. We can attempt to solve for the amplitudes using Cramer's Rule[1]. The results are

$$X_1 = \frac{\det \begin{bmatrix} 0 & -k_2 \\ 0 & -\omega^2 m_2 + k_2 \end{bmatrix}}{\det \begin{bmatrix} -\omega^2 m_1 + (k_1 + k_2) & -k_2 \\ -k_2 & -\omega^2 m_2 + k_2 \end{bmatrix}} \tag{7.20}$$

---

1 Cramer's Rule says that, given a set of linear, algebraic equations

$$[A]\{x\} = \{c\}$$

the i-th element of $\{x\}$ can be found from the ratio of two determinants. The denominator is the determinant of the coefficient matrix, $[A]$, and the numerator is the determinant of the coefficient matrix, $[A]$, modified by having the right-hand side, $\{c\}$, replace its i-th column.

and

$$X_2 = \frac{\det \begin{bmatrix} -\omega^2 m_1 + (k_1 + k_2) & 0 \\ -k_2 & 0 \end{bmatrix}}{\det \begin{bmatrix} -\omega^2 m_1 + (k_1 + k_2) & -k_2 \\ -k_2 & -\omega^2 m_2 + k_2 \end{bmatrix}} \tag{7.21}$$

Since the determinant of a matrix with a column of zeros is zero, the solutions for the amplitudes, X_1 and X_2, are

$$X_1 = \frac{0}{\det \begin{bmatrix} -\omega^2 m_1 + (k_1 + k_2) & -k_2 \\ -k_2 & -\omega^2 m_2 + k_2 \end{bmatrix}} \tag{7.22}$$

and

$$X_2 = \frac{0}{\det \begin{bmatrix} -\omega^2 m_1 + (k_1 + k_2) & -k_2 \\ -k_2 & -\omega^2 m_2 + k_2 \end{bmatrix}} \tag{7.23}$$

It appears that the solution must be $X_1 = 0$ and $X_2 = 0$, resulting in no motion of the system. This is, in fact, the solution for arbitrary values of ω. The equations of motion we are using are the free vibration equations so we are looking for the *natural motions* of the system. The question we are asking is *What would the system like to do?* and it clearly doesn't want to move at arbitrary frequencies.

There is a possibility that the amplitudes can have non-zero values and that there will be motion. Since the numerators in Equations 7.22 and 7.23 are both zero, we can consider the case where the denominators are also zero so that

$$X_1 = \frac{0}{0} \text{ and } X_2 = \frac{0}{0}$$

which are both indeterminate but not necessarily zero. Both denominators are the same so we can get this result by requiring only that

$$\det \begin{bmatrix} -\omega^2 m_1 + (k_1 + k_2) & -k_2 \\ -k_2 & -\omega^2 m_2 + k_2 \end{bmatrix} = 0 \tag{7.24}$$

Expanding the determinant gives

$$[-\omega^2 m_1 + (k_1 + k_2)][-\omega^2 m_2 + k_2] - [-k_2][-k_2] = 0 \tag{7.25}$$

which simplifies to the following fourth-order polynomial in ω

$$\omega^4 (m_1 m_2) - \omega^2 [m_2 (k_1 + k_2) + m_1 k_2] + k_1 k_2 = 0 \tag{7.26}$$

Equation 7.26 is called the *characteristic equation* for the system. There are four roots of the polynomial and they are the values of ω that will permit the system to have motion with non-zero amplitudes. We still can't say what the amplitudes are but at least we know there can be motion.

As an aside, the single degree of freedom case has many parallels with the multiple degree of freedom case. The equation of motion was

$$m\ddot{x} + kx = 0$$

and we assumed exponential or harmonic solutions with amplitude X resulting in the single equation

$$(-m\omega^2 + k)X = 0$$

This is a linear, algebraic equation with two unknowns, X and ω, so that we can't solve for both of unknowns and were content only to know the undamped natural frequency which was found from the roots of the characteristic equation to be

$$\omega_n = \pm\sqrt{\frac{k}{m}}$$

In fact, we can never know the amplitude without explicitly stating the initial conditions.

Going back to the problem at hand, we have three unknowns and two equations and we managed to find the characteristic equation that could be solved to find the natural frequencies. For the undamped case, we always get a characteristic equation that includes only even powers of ω so Equation 7.26 can actually be written as a quadratic equation in ω^2. That is

$$(\omega^2)^2 (m_1 m_2) - (\omega^2)[m_2(k_1 + k_2) + m_1 k_2] + k_1 k_2 = 0 \tag{7.27}$$

and we can use the usual formula for finding the roots of a quadratic to write

$$\omega^2 = \frac{[m_2(k_1 + k_2) + m_1 k_2] \pm \sqrt{[m_2(k_1 + k_2) + m_1 k_2]^2 - 4m_1 m_2 k_1 k_2}}{2m_1 m_2} \tag{7.28}$$

which can be slightly simplified to

$$\omega^2 = \left[\frac{(k_1 + k_2)}{2m_1} + \frac{k_2}{2m_2}\right] \pm \sqrt{\left[\frac{(k_1 + k_2)}{2m_1} + \frac{k_2}{2m_2}\right]^2 - \frac{k_1 k_2}{m_1 m_2}} \tag{7.29}$$

so that it becomes clear that there are two natural frequencies having the dimensions of $\sqrt{k/m}$. The lower natural frequency is called ω_1 and is found from

$$\omega_1^2 = \left[\frac{(k_1 + k_2)}{2m_1} + \frac{k_2}{2m_2}\right] - \sqrt{\left[\frac{(k_1 + k_2)}{2m_1} + \frac{k_2}{2m_2}\right]^2 - \frac{k_1 k_2}{m_1 m_2}} \tag{7.30}$$

and the higher natural frequency, called ω_2, is found from

$$\omega_2^2 = \left[\frac{(k_1 + k_2)}{2m_1} + \frac{k_2}{2m_2}\right] + \sqrt{\left[\frac{(k_1 + k_2)}{2m_1} + \frac{k_2}{2m_2}\right]^2 - \frac{k_1 k_2}{m_1 m_2}} \tag{7.31}$$

7.3 2DOF Undamped Free Vibrations – Mode Shapes

We were able to determine expressions for the two natural frequencies of the undamped system in the previous section. We had three unknowns and two equations so we knew we wouldn't be able to find all the unknowns but we should expect to find more than only one unknown. We therefore return to the search for amplitude information.

We were working with Equation 7.19, repeated here as Equation 7.32.

$$\begin{bmatrix} -\omega^2 m_1 + (k_1 + k_2) & -k_2 \\ -k_2 & -\omega^2 m_2 + k_2 \end{bmatrix} \begin{Bmatrix} X_1 \\ X_2 \end{Bmatrix} = \begin{Bmatrix} 0 \\ 0 \end{Bmatrix} \tag{7.32}$$

We can write the two equations in the matrix separately as

$$[-\omega^2 m_1 + (k_1 + k_2)]X_1 - k_2 X_2 = 0 \tag{7.33}$$

and

$$-k_2 X_1 + [-\omega^2 m_2 + k_2]X_2 = 0 \tag{7.34}$$

and then solve for the ratio X_1/X_2 from them.

From Equation 7.33 we get

$$\frac{X_1}{X_2} = \frac{k_2}{[-\omega^2 m_1 + (k_1 + k_2)]} \tag{7.35}$$

and, from Equation 7.34

$$\frac{X_1}{X_2} = \frac{[-\omega^2 m_2 + k_2]}{k_2} \tag{7.36}$$

Since we know the two values of ω that permit non-zero amplitudes from the previous section, we could substitute them into these equations and find the ratio of the amplitudes at each natural frequency. These ratios tell us the relative size of the amplitude of each of the masses as well as their relative phase. These are called *mode shapes* and are very useful design tools.

Equations 7.35 and 7.36 must give the same ratio or the information is not of any use to us. We can equate them to see that they do. They will be the same if

$$\frac{k_2}{[-\omega^2 m_1 + (k_1 + k_2)]} = \frac{[-\omega^2 m_2 + k_2]}{k_2} \tag{7.37}$$

Cross multiplying gives the requirement as

$$k_2 k_2 = [-\omega^2 m_1 + (k_1 + k_2)][-\omega^2 m_2 + k_2] \tag{7.38}$$

or, we must have,

$$[-\omega^2 m_1 + (k_1 + k_2)][-\omega^2 m_2 + k_2] - k_2 k_2 = 0 \tag{7.39}$$

which is characteristic equation as written in Equation 7.25. We know that the characteristic equation is satisfied by the natural frequencies, ω_1 and ω_2, so the only requirement that we have for the amplitude ratios to be the same is that the system oscillate at one or the other of its natural frequencies.

7.3.1 An Example

Consider the case where the 2DOF system has equal masses ($m_1 = m_2 = m$) and equal stiffnesses ($k_1 = k_2 = k$). The equations of motion, found by substituting into Equation 7.11, are

$$\begin{bmatrix} m & 0 \\ 0 & m \end{bmatrix} \begin{Bmatrix} \ddot{x}_1 \\ \ddot{x}_2 \end{Bmatrix} + \begin{bmatrix} 2k & -k \\ -k & k \end{bmatrix} \begin{Bmatrix} x_1 \\ x_2 \end{Bmatrix} = \begin{Bmatrix} 0 \\ 0 \end{Bmatrix} \tag{7.40}$$

Assuming harmonic motion with amplitudes X_1 and X_2 will result in algebraic equations, as follows

$$\begin{bmatrix} -\omega^2 m + 2k & -k \\ -k & -\omega^2 m + k \end{bmatrix} \begin{Bmatrix} X_1 \\ X_2 \end{Bmatrix} = \begin{Bmatrix} 0 \\ 0 \end{Bmatrix} \tag{7.41}$$

The characteristic equation is

$$(-\omega^2 m + 2k)(-\omega^2 m + k) - k^2 = 0 \tag{7.42}$$

which simplifies to

$$m^2 (\omega^2)^2 - 3km(\omega^2) + k^2 = 0 \tag{7.43}$$

The natural frequencies are then

$$\omega^2 = \frac{3km \pm \sqrt{9k^2 m^2 - 4k^2 m^2}}{2m^2} = \left(\frac{3 \pm \sqrt{5}}{2}\right)\frac{k}{m} \tag{7.44}$$

The squares of the two natural frequencies are then

$$\omega_1^2 = \left(\frac{3 - \sqrt{5}}{2}\right)\frac{k}{m} = 0.382\frac{k}{m} \tag{7.45}$$

and

$$\omega_2^2 = \left(\frac{3 + \sqrt{5}}{2}\right)\frac{k}{m} = 2.618\frac{k}{m} \tag{7.46}$$

and the two natural frequencies are

$$\omega_1 = \sqrt{0.382\frac{k}{m}} = 0.618\sqrt{\frac{k}{m}} \tag{7.47}$$

and

$$\omega_2 = \sqrt{2.618\frac{k}{m}} = 1.618\sqrt{\frac{k}{m}} \tag{7.48}$$

The mode shape corresponding to the lowest natural frequency[2] is found by substituting ω_1^2 from Equation 7.45 into Equation 7.41 to get

$$\begin{bmatrix} -0.382\frac{k}{m}m + 2k & -k \\ -k & -0.382\frac{k}{m}m + k \end{bmatrix} \begin{Bmatrix} X_1 \\ X_2 \end{Bmatrix} = \begin{Bmatrix} 0 \\ 0 \end{Bmatrix} \tag{7.49}$$

This can be tidied up by dividing throughout by k to yield

$$\begin{bmatrix} 1.618 & -1 \\ -1 & 0.618 \end{bmatrix} \begin{Bmatrix} X_1 \\ X_2 \end{Bmatrix} = \begin{Bmatrix} 0 \\ 0 \end{Bmatrix} \tag{7.50}$$

The two equations can be extracted from the matrix, giving

$$1.618X_1 - X_2 = 0 \tag{7.51}$$

and

$$-X_1 + 0.618X_2 = 0 \tag{7.52}$$

The ratio, X_1/X_2, from Equation 7.51 is

$$\left(\frac{X_1}{X_2}\right)_1 = \frac{1}{1.618} = 0.618 \tag{7.53}$$

---

2 Convention says that this is the "first" natural frequency. Natural frequencies are normally numbered from lowest to highest.

and, from Equation 7.52, is

$$\left(\frac{X_1}{X_2}\right)_1 = \frac{0.618}{1} = 0.618 \tag{7.54}$$

where the subscript "1" on the ratios reflects the fact that this result is for the first natural frequency, ω_1.

The mode shape corresponding to the first natural frequency then has the amplitude of the lower mass, X_1, being about 62% of the amplitude of the upper mass, X_2, and both of them moving in phase with each other[3].

Doing this again using the second natural frequency gives

$$\left(\frac{X_1}{X_2}\right)_2 = -1.618 \tag{7.55}$$

The second mode shape therefore has the amplitude of the lower mass being larger than the amplitude of the upper mass by about 62% and the two masses moving out of phase with each other.

Figures 7.3 and 7.4 show complete cycles of motion for the first and second mode shapes respectively. There are two types of dashed lines on each figure. The horizontal lines are the equilibrium positions of the masses. The wavy lines trace the center of each block as it moves through a cycle of motion and, therefore, represent the magnitude of the degrees of freedom.

The phase relationships between the masses can be seen clearly in the figures. The first mode has the masses moving in-phase with each other so they go up together and then down together. In the second mode, the masses are moving out-of-phase with each other so, when one goes up, the other moves down.

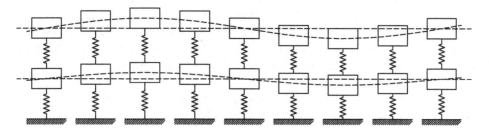

Figure 7.3 First mode shape for the 2DOF undamped system.

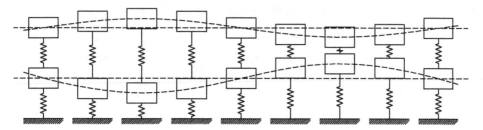

Figure 7.4 Second mode shape for the 2DOF undamped system.

3 The phase relationship is given by the sign of the ratio – positive means they are both moving in the positive directions assigned by the degrees of freedom and negative means that one is moving in the positive direction while the other is moving in the negative direction.

You can see that the spring deflections are much higher in the second mode due to the out-of-phase motion. As a rule, higher-frequency modes experience higher spring deflections. We can argue this on the basis of energy. This is an undamped system so the energy will be conserved. When the system is at maximum amplitude (i.e. at the third and seventh frames in the figures) there is no velocity (look at the slopes of the waves there) and all of the energy is potential energy stored in the springs. When the amplitudes are zero (i.e. at the first, fifth, and ninth frames) the velocity is maximum (again, look at the slopes of the waves) and the spring deflections are zero so all of the energy is kinetic energy. If we equate this kinetic energy to the potential energy at maximum amplitude we will see that the higher spring deflections mean higher potential energy and, when all of this energy is transformed to kinetic energy, we will have higher velocities. Velocity is proportional to frequency so we expect higher frequencies for the out-of-phase motions.

The amplitudes shown in the figures have been chosen so that the ratio, X_1/X_2, is correct for each mode but, other than that, the amplitudes are arbitrary. There are two reasons we can do this. Mathematically, we can't solve for X_1 and X_2 because we had two equations and three unknowns. The equations are satisfied for any X_1 and X_2 so long as they have the correct ratio. Physically, we can't know X_1 and X_2 because we never specified the initial conditions. We can therefore assume that we can choose the initial conditions that give us the amplitudes we see in the figures.

7.4 Mode Shape Descriptions

For two degree of freedom systems, it is convenient to describe mode shapes as the ratio of the amplitudes. For more than two degrees of freedom, this description becomes unwieldy because there are so many possible ratios. For example, if there are three degrees of freedom, we can write X_1/X_2 and X_1/X_3 or should we write X_2/X_3'? For even more degrees of freedom it becomes ever more difficult to decide what to do. We therefore use matrix notation to write the mode shapes.

The general procedure for finding a mode shape can be described as follows.

Given the N equations of motion for an N degree of freedom system

$$[M]\{\ddot{x}\} + [K]\{x\} = \{0\} \tag{7.56}$$

we assume harmonic solutions and derive the algebraic equations

$$[-\omega^2 M + K]\{X\} = \{0\} \tag{7.57}$$

and then develop the characteristic equation and find the N natural frequencies.

To find the i-th mode shape, we substitute natural frequency i, ω_i, into the algebraic equations to get

$$[-\omega_i^2 M + K]\{X\}_i = \{0\} \tag{7.58}$$

We know that the coefficient matrix is singular when we substitute a natural frequency into it because the natural frequencies were found by making the determinant of the coefficient matrix equal to zero. We therefore remove one equation from the set and the remaining set of N-1 equations will not be singular. We do this by assigning a value of 1 to one of the amplitudes, say X_N, so that the algebraic equations become

$$[-\omega_i^2 M + K]\begin{Bmatrix} X_1 \\ X_2 \\ \vdots \\ 1 \end{Bmatrix}_i = \{0\} \tag{7.59}$$

The N-th equation can be removed from Equation 7.59 and the N-th column, multiplied by one, can be moved to the right-hand side with a change of sign to give $(N-1)$ equations of the form

$$[-\omega_i^2 M + K]\begin{Bmatrix} X_1 \\ X_2 \\ \vdots \\ X_{N-1} \end{Bmatrix}_i = -[-\omega_i^2 M + K]\begin{Bmatrix} 0 \\ 0 \\ \vdots \\ 1 \end{Bmatrix} \tag{7.60}$$

These $(N-1)$ equations are not singular and can be solved to find the remaining elements of the mode shape, $\{X\}_i$, given that $X_N = 1$ so that the mode shape corresponding to natural frequency, ω_i becomes

$$\{X\}_i = \begin{Bmatrix} X_1 \\ X_2 \\ \vdots \\ X_{N-1} \\ 1 \end{Bmatrix}_i \tag{7.61}$$

It is important to realize that the mode shape calculated from this process is not an exact determination of the amplitude of motion of the system. It is only a representation of the relative amplitudes of the degrees of freedom. Therefore, the mode shape given in Equation 7.61 is completely scalable so that, not only is

$$\{X\}_i = \begin{Bmatrix} X_1 \\ X_2 \\ \vdots \\ X_{N-1} \\ 1 \end{Bmatrix}_i \tag{7.62}$$

a mode shape, but so are

$$\{X\}_i = \begin{Bmatrix} 10X_1 \\ 10X_2 \\ \vdots \\ 10X_{N-1} \\ 10 \end{Bmatrix}_i \tag{7.63}$$

and also

$$\{X\}_i = \begin{Bmatrix} \sqrt{2}X_1 \\ \sqrt{2}X_2 \\ \vdots \\ \sqrt{2}X_{N-1} \\ \sqrt{2} \end{Bmatrix}_i \tag{7.64}$$

We have been considering the response of the system without specifying what initial conditions caused the motion to ensue so we have determined only the *natural motions* of the system. This is what the system wants to do if disturbed away from equilibrium. Whether we are considering a 2DOF system or an NDOF system, if disturbed, the system's motion will be a superposition of its mode shapes.

7.5 Response to Initial Conditions

We start this section with a new system that we haven't considered before. Figure 7.5 shows the undamped 2DOF system that we will use as an example here.

The kinetic energy is

$$T = \frac{1}{2}m\dot{x}_1^2 + \frac{1}{2}\left(\frac{2}{3}m\right)\dot{x}_2^2 \tag{7.65}$$

The potential energy is

$$U = \frac{1}{2}kx_1^2 + \frac{1}{2}k(x_2 - x_1)^2 \tag{7.66}$$

Using Lagrange's equations, we can write

for x_1

$$\frac{\partial T}{\partial \dot{x}_1} = m\dot{x}_1 \tag{7.67}$$

$$\frac{d}{dt}\left(\frac{\partial T}{\partial \dot{x}_1}\right) = m\ddot{x}_1 \tag{7.68}$$

$$\frac{\partial T}{\partial x_1} = 0 \tag{7.69}$$

$$\frac{\partial U}{\partial x_1} = 2kx_1 - kx_2 \tag{7.70}$$

$$Q_{x_1} = 0 \tag{7.71}$$

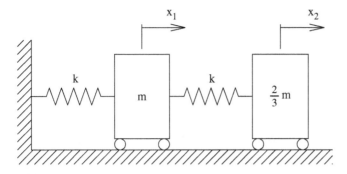

Figure 7.5 Another sample system.

and, for x_2

$$\frac{\partial T}{\partial \dot{x}_2} = \frac{2}{3}m\dot{x}_2 \qquad (7.72)$$

$$\frac{d}{dt}\left(\frac{\partial T}{\partial \dot{x}_2}\right) = \frac{2}{3}m\ddot{x}_2 \qquad (7.73)$$

$$\frac{\partial T}{\partial x_2} = 0 \qquad (7.74)$$

$$\frac{\partial U}{\partial x_2} = kx_2 - kx_1 \qquad (7.75)$$

$$Q_{x_2} = 0 \qquad (7.76)$$

The equations of motion are, therefore

$$\begin{bmatrix} m & 0 \\ 0 & \frac{2}{3}m \end{bmatrix} \begin{Bmatrix} \ddot{x}_1 \\ \ddot{x}_2 \end{Bmatrix} + \begin{bmatrix} 2k & -k \\ -k & k \end{bmatrix} \begin{Bmatrix} x_1 \\ x_2 \end{Bmatrix} = \begin{Bmatrix} 0 \\ 0 \end{Bmatrix} \qquad (7.77)$$

Assuming harmonic solutions, we get the following algebraic equations

$$\begin{bmatrix} -\omega^2 m + 2k & -k \\ -k & -\frac{2}{3}\omega^2 m + k \end{bmatrix} \begin{Bmatrix} X_1 \\ X_2 \end{Bmatrix} = \begin{Bmatrix} 0 \\ 0 \end{Bmatrix} \qquad (7.78)$$

The characteristic equation is

$$(-\omega^2 m + 2k)\left(-\frac{2}{3}\omega^2 m + k\right) - (-k)(-k) = 0 \qquad (7.79)$$

or

$$(\omega^2)^2 \left(\frac{2}{3}m^2\right) - (\omega^2)\left(\frac{7}{3}mk\right) + k^2 = 0 \qquad (7.80)$$

Without going into detail, the solutions to these equations are

$$\omega_1 = \sqrt{\frac{k}{2m}} \qquad (7.81)$$

with mode shape

$$\{X\}_1 = \begin{Bmatrix} 1 \\ \frac{3}{2} \end{Bmatrix} \qquad (7.82)$$

and

$$\omega_2 = \sqrt{\frac{3k}{m}} \tag{7.83}$$

with mode shape

$$\{X\}_2 = \begin{Bmatrix} 1 \\ -1 \end{Bmatrix} \tag{7.84}$$

Now, let us start the motion of the system in Figure 7.5 with initial conditions

$$\begin{Bmatrix} x_1(0) \\ x_2(0) \end{Bmatrix} = \begin{Bmatrix} x_{10} \\ x_{20} \end{Bmatrix} \tag{7.85}$$

and

$$\begin{Bmatrix} \dot{x}_1(0) \\ \dot{x}_2(0) \end{Bmatrix} = \begin{Bmatrix} v_{10} \\ v_{20} \end{Bmatrix} \tag{7.86}$$

The general solution is a superposition of the natural solutions so we let it be

$$\begin{Bmatrix} x_1(t) \\ x_2(t) \end{Bmatrix} = A_1 \begin{Bmatrix} 1 \\ \frac{3}{2} \end{Bmatrix} \cos \omega_1 t + A_2 \begin{Bmatrix} 1 \\ \frac{3}{2} \end{Bmatrix} \sin \omega_1 t$$

$$+ A_3 \begin{Bmatrix} 1 \\ -1 \end{Bmatrix} \cos \omega_2 t + A_4 \begin{Bmatrix} 1 \\ -1 \end{Bmatrix} \sin \omega_2 t \tag{7.87}$$

where A_1 through A_4 are unknown amplitudes to be determined from the initial conditions. The initial velocity specification requires the derivative of Equation 7.87, which is

$$\begin{Bmatrix} \dot{x}_1(t) \\ \dot{x}_2(t) \end{Bmatrix} = -\omega_1 A_1 \begin{Bmatrix} 1 \\ \frac{3}{2} \end{Bmatrix} \sin \omega_1 t + \omega_1 A_2 \begin{Bmatrix} 1 \\ \frac{3}{2} \end{Bmatrix} \cos \omega_1 t$$

$$- \omega_2 A_3 \begin{Bmatrix} 1 \\ -1 \end{Bmatrix} \sin \omega_2 t + \omega_2 A_4 \begin{Bmatrix} 1 \\ -1 \end{Bmatrix} \cos \omega_2 t \tag{7.88}$$

Substituting $t = 0$ and the initial displacements into Equation 7.87 gives

$$\begin{Bmatrix} x_{10} \\ x_{20} \end{Bmatrix} = A_1 \begin{Bmatrix} 1 \\ \frac{3}{2} \end{Bmatrix} + A_3 \begin{Bmatrix} 1 \\ -1 \end{Bmatrix} = \begin{bmatrix} 1 & 1 \\ \frac{3}{2} & -1 \end{bmatrix} \begin{Bmatrix} A_1 \\ A_3 \end{Bmatrix} \tag{7.89}$$

Using Cramer's Rule, we can write

$$A_1 = \frac{\det \begin{bmatrix} x_{10} & 1 \\ x_{20} & -1 \end{bmatrix}}{\det \begin{bmatrix} 1 & 1 \\ \frac{3}{2} & -1 \end{bmatrix}} = \frac{-x_{10} - x_{20}}{-1 - \frac{3}{2}} = \frac{x_{10} + x_{20}}{\frac{5}{2}} \tag{7.90}$$

from which

$$A_1 = \frac{2}{5}(x_{10} + x_{20}) \tag{7.91}$$

Similarly, we can find

$$A_3 = -\frac{2}{5}\left(-\frac{3}{2}x_{10} + x_{20}\right) \tag{7.92}$$

$$A_2 = \frac{2}{5}\left(\frac{v_{10} + v_{20}}{\omega_1}\right) \tag{7.93}$$

and

$$A_4 = -\frac{2}{5}\left(\frac{-\frac{3}{2}v_{10} + v_{20}}{\omega_2}\right) \tag{7.94}$$

7.6 2DOF Undamped Forced Vibrations

Consider now the same system considered in Section 7.5 with the difference that harmonic forces, $F_1(t)$ and $F_2(t)$, are applied to the masses as shown in Figure 7.6.

We can use Lagrange's Equation to derive the equations of motion as we did in Equations 7.67 through 7.76 with the only difference being that we now have generalized forces

$$Q_{x_1} = F_1(t) \tag{7.95}$$

and

$$Q_{x_2} = F_2(t) \tag{7.96}$$

so that the equations of motion are

$$\begin{bmatrix} m & 0 \\ 0 & \frac{2}{3}m \end{bmatrix} \begin{Bmatrix} \ddot{x}_1 \\ \ddot{x}_2 \end{Bmatrix} + \begin{bmatrix} 2k & -k \\ -k & k \end{bmatrix} \begin{Bmatrix} x_1 \\ x_2 \end{Bmatrix} = \begin{Bmatrix} F_1(t) \\ F_2(t) \end{Bmatrix} \tag{7.97}$$

We consider the steady state response of the system to harmonic forces with forcing frequency, ω. To that end, let

$$\begin{Bmatrix} F_1(t) \\ F_2(t) \end{Bmatrix} = \begin{Bmatrix} F_1 \\ F_2 \end{Bmatrix} e^{i\omega t} \tag{7.98}$$

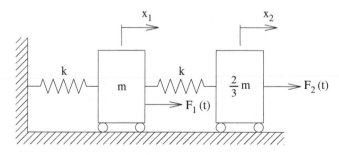

Figure 7.6 A 2DOF undamped, forced, system.

Clearly, the particular solution to Equation 7.97 must then be a harmonic oscillation at the forcing frequency so we let

$$\left\{ \begin{matrix} x_1(t) \\ x_2(t) \end{matrix} \right\} = \left\{ \begin{matrix} X_1 \\ X_2 \end{matrix} \right\} e^{i\omega t} \tag{7.99}$$

and then differentiate it twice and substitute it and the forcing function (Equation 7.98) into Equation 7.97 to get

$$\begin{bmatrix} m & 0 \\ 0 & \frac{2}{3}m \end{bmatrix} \left\{ \begin{matrix} -\omega^2 X_1 \\ -\omega^2 X_2 \end{matrix} \right\} e^{i\omega t} + \begin{bmatrix} 2k & -k \\ -k & k \end{bmatrix} \left\{ \begin{matrix} X_1 \\ X_2 \end{matrix} \right\} e^{i\omega t} = \left\{ \begin{matrix} F_1 \\ F_2 \end{matrix} \right\} e^{i\omega t} \tag{7.100}$$

Canceling out the $e^{i\omega t}$ terms, moving $-\omega^2$ into the mass matrix, and adding the coefficient matrices together gives the following set of linear, algebraic equations.

$$\begin{bmatrix} -m\omega^2 + 2k & -k \\ -k & -\frac{2}{3}m\omega^2 + k \end{bmatrix} \left\{ \begin{matrix} X_1 \\ X_2 \end{matrix} \right\} = \left\{ \begin{matrix} F_1 \\ F_2 \end{matrix} \right\} \tag{7.101}$$

Equation 7.101 is a set of linear algebraic equations that can readily be solved using Cramer's Rule. The expressions for the amplitudes, X_1 and X_2, are

$$X_1 = \frac{\det \begin{bmatrix} F_1 & -k \\ F_2 & -\frac{2}{3}m\omega^2 + k \end{bmatrix}}{\det \begin{bmatrix} -m\omega^2 + 2k & -k \\ -k & -\frac{2}{3}m\omega^2 + k \end{bmatrix}} \tag{7.102}$$

and

$$X_2 = \frac{\det \begin{bmatrix} -m\omega^2 + 2k & F_1 \\ -k & F_2 \end{bmatrix}}{\det \begin{bmatrix} -m\omega^2 + 2k & -k \\ -k & -\frac{2}{3}m\omega^2 + k \end{bmatrix}} \tag{7.103}$$

Given parameter values, the amplitudes can be calculated and plotted as a function of forcing frequency. Note that the denominators in Equations 7.102 and 7.103 are the characteristic polynomial for the system. This will go to zero twice, once at each natural frequency, causing both amplitudes to become very large when the system is forced at or near any of its natural frequencies.

7.7 Vibration Absorbers

Consider the system shown in Figure 7.7 which is a model of a piece of equipment for which you are responsible. This is a 1DOF undamped system subject to harmonic base motions. The equation of motion, for $Y(t) = Y \sin \omega t$, is

$$m\ddot{x} + kx = kY \sin \omega t \tag{7.104}$$

and the amplitude of motion, X, can be written as

$$X = \frac{kY}{-m\omega^2 + k} \tag{7.105}$$

Figure 7.7 A model of a machine experiencing large amplitudes.

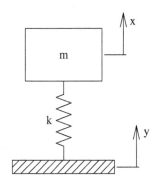

Now consider the case where the forcing frequency, ω, is near the natural frequency of the system so that the motions are larger than you would wish. One method of controlling the amplitude of motion is to employ a *vibration absorber*. To do this, you mount another mass-spring system to the machine, resulting in a 2DOF model as shown in Figure 7.8. The vibration absorber has mass m_a and stiffness k_a, as shown.

The equations of motion for the new system are

$$
\begin{bmatrix} m & 0 \\ 0 & m_a \end{bmatrix} \begin{Bmatrix} \ddot{x} \\ \ddot{x}_a \end{Bmatrix} + \begin{bmatrix} k + k_a & -k_a \\ -k_a & k_a \end{bmatrix} \begin{Bmatrix} x \\ x_a \end{Bmatrix} = \begin{Bmatrix} k \\ 0 \end{Bmatrix} Y \sin \omega t \tag{7.106}
$$

Assuming harmonic response at the forcing frequency, the amplitudes are

$$
X = \frac{\det \begin{bmatrix} kY & -k_a \\ 0 & -m_a \omega^2 + k_a \end{bmatrix}}{\det \begin{bmatrix} -m\omega^2 + (k + k_a) & -k_a \\ -k_a & -m_a \omega^2 + k_a \end{bmatrix}} \tag{7.107}
$$

Figure 7.8 The machine with the vibration absorber mounted.

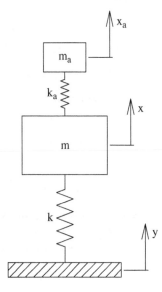

and

$$X_a = \frac{\det \begin{bmatrix} -m\omega^2 + (k + k_a) & kY \\ -k_a & 0 \end{bmatrix}}{\det \begin{bmatrix} -m\omega^2 + (k + k_a) & -k_a \\ -k_a & -m_a\omega^2 + k_a \end{bmatrix}} \tag{7.108}$$

The denominators of these two expressions are the characteristic polynomial of the system, as expected. The numerators are the interesting parts. From Equation 7.107, we can find that the numerator for the amplitude of the machine, X, is

$$kY(-m_a\omega^2 + k_a) \tag{7.109}$$

We can set this to zero by choosing k_a and m_a so that

$$\sqrt{\frac{k_a}{m_a}} = \omega \quad \text{the forcing frequency} \tag{7.110}$$

The vibration absorber is a device that is tuned so that its natural frequency is equal to the frequency that is disturbing the system. It responds as if it is a 1DOF system at this frequency and absorbs all of the energy coming into the system so that the original structure stops moving.

The design of vibration absorbers is a much bigger topic than can be addressed here. There are methods for optimizing the mass and damping in the absorbers for specific applications that are covered in the specialized literature on the subject.

7.8 The Method of Normal Modes

Consider a forced, undamped, system with N degrees of freedom. The equations of motion would be

$$[M]\{\ddot{x}\} + [K]\{x\} = \{f(t)\} \tag{7.111}$$

We first solve the free vibration problem for this system

$$[M]\{\ddot{x}\} + [K]\{x\} = \{0\} \tag{7.112}$$

so that we know the natural frequencies and their corresponding mode shapes. Therefore, let

$$\{x(t)\} = \{X\}\sin\omega t \tag{7.113}$$

resulting in

$$-\omega^2[M]\{X\} + [K]\{X\} = \{0\} \tag{7.114}$$

from which we can find the N natural frequencies, $\omega_i, i = 1, N$, and the N corresponding mode shapes, $\{X\}_i, i = 1, N$.

In fact, the problems we are talking about here are *eigenvalue* problems where the natural frequencies are the *eigenvalues* and the mode shapes are the *eigenvectors*. Eigenvectors have interesting properties that we are going to use to help us with the forced vibration problem we are considering.

The first property is *orthogonality*. We can show that the eigenvectors are orthogonal with respect to the mass matrix and the stiffness matrix so long as both matrices are symmetric. Deriving the equations of motion from Lagrange's Equation guarantees symmetric mass and stiffness matrices so the restriction is not a problem.

The proof goes as follows. Say we have found two natural frequencies, ω_i and ω_j, and their corresponding eigenvectors, $\{X\}_i$ and $\{X\}_j$. Both of these are solutions to Equation 7.114, so we can use it to write two identities as

$$\omega_i^2 [M]\{X\}_i = [K]\{X\}_i \tag{7.115}$$

and

$$\omega_j^2 [M]\{X\}_j = [K]\{X\}_j \tag{7.116}$$

Now, premultiply[4] Equation 7.115 by $\{X\}_j^T$ (the transpose of $\{X\}_j$) and Equation 7.116 by $\{X\}_i^T$, giving

$$\omega_i^2 \{X\}_j^T [M]\{X\}_i = \{X\}_j^T [K]\{X\}_i \tag{7.117}$$

and

$$\omega_j^2 \{X\}_i^T [M]\{X\}_j = \{X\}_i^T [K]\{X\}_j \tag{7.118}$$

Then, transpose Equation 7.117 to get[5]

$$\omega_i^2 \{X\}_i^T [M]^T \{X\}_j = \{X\}_i^T [K]^T \{X\}_j \tag{7.119}$$

Since $[M]$ and $[K]$ are symmetric matrices, we can write

$$[M]^T = [M] \quad \text{and} \quad [K]^T = [K] \tag{7.120}$$

so that Equation 7.119 can be written as

$$\omega_i^2 \{X\}_i^T [M]\{X\}_j = \{X\}_i^T [K]\{X\}_j \tag{7.121}$$

We can subtract Equation 7.121 from Equation 7.118 to get another identity

$$(\omega_j^2 - \omega_i^2)\{X\}_i^T [M]\{X\}_j = 0 \tag{7.122}$$

---

4 The results of matrix multiplication are dependent on the order of multiplication. *Premultiply* means to multiply on the left. The reason for the multiplication at this point is that it changes the expressions we are working with into scalars so we can eventually add and subtract them. The dimensions we have are: $[M]$ and $[K]$ are $N \times N$ square matrices; $\{X\}_i$ and $\{X\}_j$ are $N \times 1$ column vectors; and $\{X\}_i^T$ and $\{X\}_j^T$ are $1 \times N$ row vectors. Products such as $\{X\}_i^T[M]\{X\}_j$ and $\{X\}_i^T[K]\{X\}_j$ therefore have dimensions

$$(1 \times N) \times (N \times N) \times (N \times 1) = (1 \times 1)$$

so they are scalars.

5 The rule for transposing products of matrices is that the transpose of the product is equal to the product of the transposed matrices in reverse order. That is

$$([A][B][C])^T = [C]^T [B]^T [A]^T$$

Scalars such as ω_i^2 can be placed anywhere in the product.

Equation 7.122 can be considered for two cases. First we consider the case where $i \neq j$ so we are considering two modes with different natural frequencies, $\omega_i \neq \omega_j$. In this case, Equation 7.122 can only be satisfied if

$$\{X\}_i^T [M] \{X\}_j = 0 \qquad (7.123)$$

The second case is where we use the same natural frequency and mode shape twice. That is $i = j$ and $\omega_i = \omega_j$. In this case, $(\omega_j^2 - \omega_i^2) = (\omega_i^2 - \omega_i^2) = 0$ and Equation 7.122 is automatically satisfied. Without going into detail, it can be shown that, for the symmetric, positive definite mass matrix, $[M]$,

$$\{X\}_i^T [M] \{X\}_i \neq 0 \qquad (7.124)$$

unless $\{X\}_i = \{0\}$ and we have dismissed this as a solution because it means there is no motion.

The conditions expressed in Equations 7.123 and 7.124 are a statement of orthogonality . We can say that the mode shapes are orthogonal with respect to the mass matrix. Furthermore, we can go back to Equation 7.121 and recognize that the mode shapes are also orthogonal with respect to the stiffness matrix since $\{X\}_i^T [K] \{X\}_j$ is just equal to $\{X\}_i^T [M] \{X\}_j$ multiplied by a scalar, ω_i^2.

The next step is to *normalize* the mode shapes with respect to the mass matrix. Remembering that the mode shapes can be scaled arbitrarily so long as the elements retain the same relationships with respect to each other, allows us to say that, if $\{X\}_i$ is a mode shape, then so is $\{fX\}_i$ where f is a scaling factor operating on all of the elements in the mode shape.

We want to scale (normalize) the mode shapes so that

$$\{X\}_i^T [M] \{X\}_i = 1 \qquad (7.125)$$

The procedure is as follows.

- Find all the $\{X\}_i$ for the system. The numerical routines used will scale them arbitrarily.
- For each $i = 1, N$, calculate $\{X\}_i^T [M] \{X\}_i = b_i$ where b_i is a number not equal to one.
- Assume that you have scaled the mode shape by a factor f_i so that $\{f_i X\}_i$ is normalized. In that case, you would have

$$\{f_i X\}_i^T [M] \{f_i X\}_i = 1$$

or, factoring out f_i twice,

$$f_i^2 \{X\}_i^T [M]\{X\}_i = 1$$

We know that $\{X\}_i^T [M] \{X\}_i$ is equal to b_i so, we have

$$f_i^2 b_i = 1$$

which gives the scaling factor

$$f_i = \sqrt{\frac{1}{b_i}}$$

- Once all of the mode shapes have been scaled, the entire set is normalized with respect to the mass matrix

We can now look back at Equation 7.121 and see what $\{X\}_i^T [K] \{X\}_i$ is equal to.

Substituting $\{X\}_i^T [M] \{X\}_i = 1$ gives

$$\omega_i^2 = \{X\}_i^T [K] \{X\}_i \tag{7.126}$$

so that each of these products gives the natural frequency squared[6].

The next step is to define the *modal matrix*, $[U]$, which is an $N \times N$ matrix whose columns are the normalized mode shapes.

$$[U] = [\{X\}_1 \{X\}_2 \cdots \{X\}_N] \tag{7.127}$$

The modal matrix has the the following multiplicative properties.

- Since $\{X\}_i^T [M] \{X\}_j = 0$ for $i \neq j$ and $\{X\}_i^T [M] \{X\}_j = 1$ for $i = j$

$$[U]^T [M] [U] = [I] \tag{7.128}$$

where $[I]$ is the identity matrix.
- Since $\{X\}_i^T [K] \{X\}_j = 0$ for $i \neq j$ and $\{X\}_i^T [K] \{X\}_j = \omega_i^2$ for $i = j$

$$[U]^T [K] [U] = [D(\omega)] \tag{7.129}$$

where $[D(\omega)]$ is a diagonal matrix having the natural frequencies squared along the diagonal.

At this point we introduce the *expansion theorem* that says we can expand any vector of length N in terms of the N eigenvectors or mode shapes that we are working with. Let there be a vector $\{v\}$. The expansion theorem says that we can write

$$\{v\} = c_1 \{X\}_1 + c_2 \{X\}_2 + \cdots + c_N \{X\}_N \tag{7.130}$$

where the c_i are scalar coefficients.

To show that this is the case, we write Equation 7.130 in summation notation and premultiply by the mass matrix, $[M]$, to get

$$[M] \{v\} = \sum_{i=1}^{N} c_i [M] \{X\}_i \tag{7.131}$$

Then we choose one mode shape, say $\{X\}_p$ where p is any number from 1 to N, and premultiply by its transpose.

$$\{X\}_p^T [M] \{v\} = \sum_{i=1}^{N} c_i \{X\}_p^T [M] \{X\}_i \tag{7.132}$$

---

6 This is analogous to the SDOF systems where we looked at the equation of motion

$$m\ddot{x} + kx = 0$$

and divided throughout my m to get

$$\ddot{x} + \frac{k}{m} x = 0$$

and then recognized that

$$\frac{k}{m} = \omega_n^2$$

There are N terms in the summation and all but one of them is zero by orthogonality. The only non-zero term is

$$c_p\{X\}_p^T[M]\{X\}_p$$

and this is equal to c_p since $\{X\}_p^T[M]\{X\}_p = 1$. We could repeat this for all the modes and generate a set of coefficients $c_i, i = 1, N$ that expands $\{v\}$ in terms of the mode shapes.

Armed with these new tools, we finally return to the equations of motion for the forced system as initially presented in Equation 7.111 and repeated here as Equation 7.133.

$$[M]\{\ddot{x}\} + [K]\{x\} = \{f(t)\} \tag{7.133}$$

We can use the expansion theorem to expand the vector of degrees of freedom $\{x(t)\}$ in terms of the modes, yielding

$$\{x(t)\} = \sum_{i=1}^{N} c_i(t)\{X\}_i \tag{7.134}$$

where the N coefficients $c_i(t)$ are time-varying. We can use the modal matrix, $[U]$, to recast Equation 7.134 into[7]

$$\{x(t)\} = [U]\{c(t)\} \tag{7.135}$$

Equation 7.135 can be differentiated twice to give

$$\{\ddot{x}(t)\} = [U]\{\ddot{c}(t)\} \tag{7.136}$$

Substitution into the equation of motion gives

$$[M][U]\{\ddot{c}(t)\} + [K][U]\{c(t)\} = \{f(t)\} \tag{7.137}$$

which can then be premultiplied by $[U]^T$ to give

$$[U]^T[M][U]\{\ddot{c}(t)\} + [U]^T[K][U]\{c(t)\} = [U]^T\{f(t)\} \tag{7.138}$$

Substituting the results from Equations 7.128 and 7.129 shows the advantage of using the Method of Normal Modes because the equations of motion are *decoupled* and become

$$[I]\{\ddot{c}(t)\} + [D(\omega)]\{c(t)\} = [U]^T\{f(t)\} \tag{7.139}$$

Decoupled means that the degrees of freedom appear in their own equations and the off-diagonal terms in the coefficient matrices that mean we need to consider all the degrees of freedom at the same time have been removed. We now have N independent SDOF systems to consider.

---

7 This may not be immediately clear so consider a 2DOF example as follows. For 2DOF, Equation 7.134 can be written as

$$\left\{ \begin{array}{c} x_1(t) \\ x_2(t) \end{array} \right\} = c_1(t) \left\{ \begin{array}{c} X_1 \\ X_2 \end{array} \right\}_1 + c_2(t) \left\{ \begin{array}{c} X_1 \\ X_2 \end{array} \right\}_2$$

which gives the same result as

$$\left\{ \begin{array}{c} x_1(t) \\ x_2(t) \end{array} \right\} = \left[\left\{ \begin{array}{c} X_1 \\ X_2 \end{array} \right\}_1 \left\{ \begin{array}{c} X_1 \\ X_2 \end{array} \right\}_2 \right] \left\{ \begin{array}{c} c_1(t) \\ c_2(t) \end{array} \right\}$$

which is what Equation 7.135 says.

The initial conditions for $c(t)$ and $\dot{c}(t)$ are derived from the physical initial conditions as follows. Premultiply Equation 7.135 by the mass matrix to get

$$[M]\{x(t)\} = [M][U]\{c(t)\} \tag{7.140}$$

and then by the transpose of the modal matrix

$$[U]^T[M]\{x(t)\} = [U]^T[M][U]\{c(t)\} \tag{7.141}$$

and then recognize from Equation 7.128 that $[U]^T[M][U]$ is the identity matrix so we can write

$$\{c(t)\} = [U]^T[M]\{x(t)\} \tag{7.142}$$

Differentiate Equation 7.142 to get $\dot{c}(t)$ as

$$\{\dot{c}(t)\} = [U]^T[M]\{\dot{x}(t)\} \tag{7.143}$$

Finally, apply Equations 7.142 and 7.143 at $t = 0$ to get the initial conditions

$$\{c(0)\} = [U]^T[M]\{x(0)\} \tag{7.144}$$

and

$$\{\dot{c}(0)\} = [U]^T[M]\{\dot{x}(0)\} \tag{7.145}$$

7.9 The Cart and Pendulum Example

At this point, we need an example for the Method of Normal Modes and an exercise in deriving the linearized equations of motion for a more complex system than we have seen before is warranted for that purpose. It is also important that we don't fall into the habit of thinking all vibrational systems are made up of square blocks supported by coil springs as we often see as the model used for discussing techniques in vibrational analysis.

Figure 7.9 shows a cart of total mass, m, that is able to move freely on a horizontal surface. There is a single elastic element with stiffness, k, attaching the cart to ground. Inside the cart is a simple pendulum with length, ℓ, and mass, m. The cart has a horizontal force, $F \sin \omega t$, applied to it as shown. This system has only two degrees of freedom, x and θ, but one is translational while the other is rotational. This will force us to deal with motion of points in the system that arise from the combination of two different types of coordinates. In addition, the x DOF is *absolute* in that it is measured with respect to a fixed equilibrium position whereas the θ DOF is *relative* because it is measured with respect to the moving cart.

The initial goal is to derive the equations of motion for the system. This will be followed by determination of the natural frequencies and mode shapes and, finally, the time domain response will be determined using the Method of Normal Modes described in Section 7.8.

When deriving the equations of motion, we must remember that the velocities and accelerations used must be *absolute* and that θ specifies the position of the pendulum bob relative to the cart so its derivatives give relative velocity and acceleration.

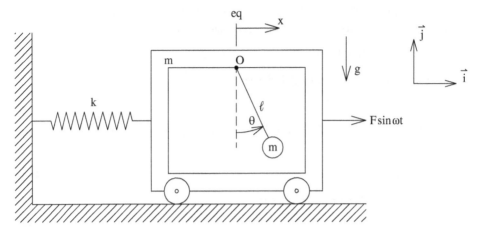

Figure 7.9 The cart and pendulum system.

7.9.1 Modeling the System – Two Ways

Here, we attack the problem of finding the linearized equations of motion in two different ways. First, we include all the nonlinear terms in a formal kinematic analysis and linearize about equilibrium at the end. We do this twice, once with Newton's Laws and then again using Lagrange's Equation. This method always works but is somewhat labor-intensive.

The second approach uses a direct linear formulation by working with informal, linearized, kinematic expressions derived graphically. This is riskier in that terms might be lost but involves much less work.

7.9.1.1 Kinematics

The first step in any dynamic analysis involves the derivation of kinematic expressions that satisfy the constraints on the motion. In this case the constraints are that the cart can only move horizontally and that the point where the pendulum is connected to the cart, point O, must have the same motion on both bodies. Notice that these constraints have already been recognized when the degrees of freedom were chosen. Now they need to be accounted for in the kinematic expressions.

In the formal approach we write the absolute velocity of the cart

$$\vec{v}_O = \dot{x}\vec{i} \tag{7.146}$$

and then differentiate it to get the absolute acceleration of the cart

$$\vec{a}_O = \ddot{x}\vec{i} \tag{7.147}$$

Next, we write the position of the pendulum bob relative to the point of connection of the pendulum to the cart and differentiate it twice, once to get the velocity of the bob relative to the cart and then to get the acceleration of the bob relative to the cart.

$$\vec{p}_{m/O} = \ell \sin \theta \vec{i} - \ell \cos \theta \vec{j} \tag{7.148}$$

$$\vec{v}_{m/O} = \ell \dot{\theta} \cos \theta \vec{i} + \ell \dot{\theta} \sin \theta \vec{j} \tag{7.149}$$

$$\vec{a}_{m/O} = (\ell \ddot{\theta} \cos \theta - \ell \dot{\theta}^2 \sin \theta)\vec{i} + (\ell \ddot{\theta} \sin \theta + \ell \dot{\theta}^2 \cos \theta)\vec{j} \tag{7.150}$$

Now, we add the velocity and acceleration of the bob relative to the cart to the absolute velocity and acceleration of the cart respectively to get the absolute velocity and acceleration of the pendulum bob

$$\vec{v}_m = \vec{v}_O + \vec{v}_{m/O} = (\dot{x} + \ell\dot{\theta}\cos\theta)\vec{i} + \ell\dot{\theta}\sin\theta\vec{j} \tag{7.151}$$

$$\vec{a}_m = \vec{a}_O + \vec{a}_{m/O} = (\ddot{x} + \ell\ddot{\theta}\cos\theta - \ell\dot{\theta}^2\sin\theta)\vec{i} + (\ell\ddot{\theta}\sin\theta + \ell\dot{\theta}^2\cos\theta)\vec{j} \tag{7.152}$$

The informal approach goes as follows, where we talk about horizontal and vertical components rather than using the unit vectors shown on Figure 7.9. For the cart, the absolute velocity is $\dot{x}$ and the absolute acceleration is $\ddot{x}$, both to the right. The pendulum bob, relative to the cart, has a tangential (i.e. perpendicular to the cord) velocity equal to $\ell\dot{\theta}$ to the right and upward. The acceleration of the bob relative to the cart comprises two components: (1) a centripetal component of $\ell\dot{\theta}^2$ pointed from m towards O, and (2) a tangential component of $\ell\ddot{\theta}$ perpendicular to the cord and tending to the right and upward.

We note that the centripetal term is nonlinear since it involves $\dot{\theta}^2$ so, in a linear sense, is negligibly small and can be dropped from further analysis. We also note that the vertical components of the tangential velocity and acceleration of the bob involve $\sin\theta$ (i.e. the vertical velocity of the bob relative to the cart is $\ell\dot{\theta}\sin\theta \approx \ell\dot{\theta}\theta$ and the vertical acceleration is $\ell\ddot{\theta}\sin\theta \approx \ell\ddot{\theta}\theta$). Both of these terms are nonlinear, and therefore negligible, because of the $\dot{\theta}\theta$ and $\ddot{\theta}\theta$ products. The horizontal components of both velocity and acceleration have $\cos\theta \approx 1$ terms in them and will be retained. The result of all this is that the linearized absolute velocity of the bob is $\dot{x} + \ell\dot{\theta}$ to the right and its linearized absolute acceleration is $\ddot{x} + \ell\ddot{\theta}$, also to the right.

All of the discussion in the previous paragraph was only to point out the process whereby we arrive at the final result. In practice, we would simply look at the drawing of the system and note that the absolute velocity and acceleration of the cart are $\dot{x}$ and $\ddot{x}$ to the right. In addition to that, the pendulum bob, when hanging straight down in equilibrium, adds a relative velocity of $\ell\dot{\theta}$ and a relative acceleration of $\ell\ddot{\theta}$, both to the right, as previously stated.

7.9.1.2 Newton's Laws

Figure 7.10 shows free body diagrams of the cart and of the pendulum bob. There are three force balance equations to write: horizontal on the cart; horizontal on the bob; vertical on the bob. Once these are written, they will be combined to eliminate the internal constraint force in the cable and two equations will remain. These will be the equations of motion.

The force balance equations are, on the cart

$$+ \rightarrow \sum F : F\sin\omega t + T\sin\theta - kx = ma_{O_{\vec{i}}} \tag{7.153}$$

and, on the pendulum bob

$$+ \rightarrow \sum F : -T\sin\theta = ma_{m_{\vec{i}}} \tag{7.154}$$

$$+ \uparrow \sum F : T\cos\theta - mg = ma_{m_{\vec{j}}} \tag{7.155}$$

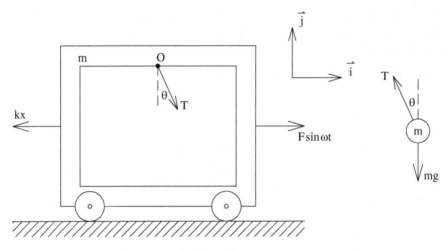

Figure 7.10 The cart and pendulum system – FBD.

Substituting the formal kinematic expressions from Equations 7.147 and 7.152 give

$$F \sin \omega t + T \sin \theta - kx = m\ddot{x} \tag{7.156}$$

$$- T \sin \theta = m(\ddot{x} + \ell \ddot{\theta} \cos \theta - \ell \dot{\theta}^2 \sin \theta) \tag{7.157}$$

$$T \cos \theta - mg = m(\ell \ddot{\theta} \sin \theta + \ell \dot{\theta}^2 \cos \theta) \tag{7.158}$$

Equations 7.156 and 7.157 can be added to eliminate $T \sin \theta$ with the result, after some simplification, that

$$2m\ddot{x} + m\ell \ddot{\theta} \cos \theta - m\ell \dot{\theta}^2 \sin \theta + kx = F \sin \omega t \tag{7.159}$$

Multiplying Equation 7.157 by $\cos \theta$ everywhere and adding the result to $\sin \theta$ multiplied by Equation 7.158 gives, again after some simplification[8]

$$m\ddot{x} \cos \theta + m\ell \ddot{\theta} + mg \sin \theta = 0 \tag{7.160}$$

Equations 7.159 and 7.160 are the nonlinear equations of motion for the system. For small motions about the equilibrium state ($x = 0$ and $\theta = 0$), we get $\sin \theta \approx \theta$ and $\cos \theta \approx 1$ so that the linearized equations of motion are

$$\begin{bmatrix} 2m & m\ell \\ m & m\ell \end{bmatrix} \begin{Bmatrix} \ddot{x} \\ \ddot{\theta} \end{Bmatrix} + \begin{bmatrix} k & 0 \\ 0 & mg \end{bmatrix} \begin{Bmatrix} x \\ \theta \end{Bmatrix} = \begin{Bmatrix} F \sin \omega t \\ 0 \end{Bmatrix} \tag{7.161}$$

Using the informal kinematics from Subsection 7.9.1.1 in the force balance equations (Equations 7.153 to 7.155) and linearizing simultaneously, leads to

$$F \sin \omega t + T\theta - kx = m\ddot{x} \tag{7.162}$$

$$- T\theta = m(\ddot{x} + \ell \ddot{\theta}) \tag{7.163}$$

$$T - mg = 0 \tag{7.164}$$

---

8 The constraint force, T, can be found by multiplying Equation 7.157 by $\sin \theta$, multiplying Equation 7.158 by $\cos \theta$ and subtracting the two results to get $T = mg \cos \theta - m\ddot{x} \sin \theta + m\ell \dot{\theta}^2$. Notice the two nonlinear terms that influence the tension force. They can't be found by any of the other methods presented here.

Adding Equations 7.162 and 7.163 eliminates T immediately, leaving

$$F \sin \omega t = 2m\ddot{x} + m\ell\ddot{\theta} + kx \tag{7.165}$$

Then, $T = mg$ from Equation 7.164 can be substituted directly into Equation 7.163 to give

$$-mg\theta = m\ddot{x} + m\ell\ddot{\theta} \tag{7.166}$$

Equations 7.165 and 7.166 are exactly the same equations of motion as are in Equation 7.161.

7.9.1.3 Lagrange's Equation

To derive the equations of motion using Lagrange's Equation, we need expressions for the kinetic energy, T, and the potential energy U.

Using the formal kinematics method and the general expression for the kinetic energy of 2D rigid bodies,

$$T = \frac{1}{2}mv_G^2 + \frac{1}{2}I_G\omega^2$$

where m is the mass of the body, v_G is the magnitude of the absolute velocity of the center of mass of the body, I_G is the moment of inertia of the body about its center of mass, and ω is the absolute angular velocity of the body, we find

$$T = \frac{1}{2}m\dot{x}^2 + \frac{1}{2}m[(\dot{x} + \ell\dot{\theta}\cos\theta)^2 + (\ell\dot{\theta}\sin\theta)^2] \tag{7.167}$$

which simplifies to

$$T = m\dot{x}^2 + m\ell\dot{x}\dot{\theta}\cos\theta + \frac{1}{2}m\ell^2\dot{\theta}^2 \tag{7.168}$$

The potential energy in this system is due to both gravitational and elastic effects. Taking the datum for potential energy to be at the level of the connection of the pendulum to the cart (point O), we can write[9]

$$U = \frac{1}{2}kx^2 - mg\ell\cos\theta \tag{7.169}$$

The terms we need for Lagrange's Equation are, first for x,

$$\frac{\partial T}{\partial \dot{x}} = 2m\dot{x} + m\ell\dot{\theta}\cos\theta$$

$$\frac{d}{dt}\left(\frac{\partial T}{\partial \dot{x}}\right) = 2m\ddot{x} + m\ell\ddot{\theta}\cos\theta - m\ell\dot{\theta}^2\sin\theta$$

$$\frac{\partial T}{\partial x} = 0$$

$$\frac{\partial U}{\partial x} = kx$$

$$Q_x \delta x = (F\sin\omega t)\delta x \Rightarrow Q_x = F\sin\omega t \tag{7.170}$$

---

9 Since Lagrange's Equation includes derivatives (i.e. rates of *change*) of the potential energy with respect to generalized coordinates, we need only be concerned with changes in potential energy. Thus we define a constant level as a datum for gravitational potential energy, where the actual level is of no importance so long as it never changes. Looking for potential energy changes also allows us to leave out the gravitational potential energy of bodies, like the cart, that never move vertically.

and, for θ,

$$\frac{\partial T}{\partial \dot{\theta}} = m\dot{x}\ell \cos\theta + m\ell^2\dot{\theta}$$

$$\frac{d}{dt}\left(\frac{\partial T}{\partial \dot{\theta}}\right) = m\ddot{x}\ell \cos\theta - m\dot{x}\ell\dot{\theta} \sin\theta + m\ell^2\ddot{\theta}$$

$$\frac{\partial T}{\partial \theta} = -m\dot{x}\ell\dot{\theta} \sin\theta$$

$$\frac{\partial U}{\partial \theta} = mg\ell \sin\theta$$

$$Q_\theta = 0 \tag{7.171}$$

Linearizing and substituting into Lagrange's Equation yields

$$\begin{bmatrix} 2m & m\ell \\ m\ell & m\ell^2 \end{bmatrix} \left\{ \begin{matrix} \ddot{x} \\ \ddot{\theta} \end{matrix} \right\} + \begin{bmatrix} k & 0 \\ 0 & mg\ell \end{bmatrix} \left\{ \begin{matrix} x \\ \theta \end{matrix} \right\} = \left\{ \begin{matrix} F\sin\omega t \\ 0 \end{matrix} \right\} \tag{7.172}$$

Note that Lagrange always gives symmetric mass and stiffness matrices unlike Newton (see Equation 7.161) where the final result will depend on how the force balance equations were combined in order to remove internal constraint forces from the equations of motion. If you multiply the second equation in Equation 7.161 by ℓ, the two sets of equations are identical.

We can also use the informal, linear, kinematics expressions derived in Subsection 7.9.1.1 with Lagrange's Equation. We found there that the absolute velocity of the pendulum bob was, keeping linear terms only, $\dot{x} + \ell\dot{\theta}$ to the right. With the addition of the cart, this gives the total kinetic energy as

$$T = \frac{1}{2}m\dot{x}^2 + \frac{1}{2}m(\dot{x} + \ell\dot{\theta})^2 = m\dot{x}^2 + m\ell\dot{x}\dot{\theta} + \frac{1}{2}m\ell^2\dot{\theta}^2 \tag{7.173}$$

The potential energy, placing the datum at the level of the connection of the pendulum to the cart as we did before, is

$$U = \frac{1}{2}kx^2 - mg\ell \cos\theta \approx \frac{1}{2}kx^2 - mg\ell \tag{7.174}$$

We have a problem here. If the gravitational potential energy is a constant, there will be no gravitational terms in the equations of motion because all derivatives of the gravitational potential energy will be zero. We have, in fact, linearized too soon in this case. In order to be able to differentiate the potential energy, we must either keep it nonlinear by retaining the $\cos\theta$ term until after the differentiation or we could keep the first two terms of the Maclaurin series for the cosine (i.e. $\cos\theta \approx 1 - \theta^2/2!$) so that the differentiation leads directly to the linear expression.

Using the informal kinematics, the terms we need for Lagrange's Equation are, for x,

$$\frac{\partial T}{\partial \dot{x}} = 2m\dot{x} + m\ell\dot{\theta}$$

$$\frac{d}{dt}\left(\frac{\partial T}{\partial \dot{x}}\right) = 2m\ddot{x} + m\ell\ddot{\theta}$$

$$\frac{\partial T}{\partial x} = 0$$

$$\frac{\partial U}{\partial x} = kx$$

$$Q_x \delta x = (F \sin \omega t)\delta x \Rightarrow Q_x = F \sin \omega t \tag{7.175}$$

and, for θ,

$$\frac{\partial T}{\partial \dot{\theta}} = m\dot{x}\ell + m\ell^2\dot{\theta}$$

$$\frac{d}{dt}\left(\frac{\partial T}{\partial \dot{\theta}}\right) = m\ddot{x}\ell + m\ell^2\ddot{\theta}$$

$$\frac{\partial T}{\partial \theta} = 0$$

$$\frac{\partial U}{\partial \theta} = mg\ell \sin\theta \approx mg\ell\theta$$

$$Q_\theta = 0 \tag{7.176}$$

Substituting into Lagrange's equation gives

$$\begin{bmatrix} 2m & m\ell \\ m\ell & m\ell^2 \end{bmatrix}\begin{Bmatrix} \ddot{x} \\ \ddot{\theta} \end{Bmatrix} + \begin{bmatrix} k & 0 \\ 0 & mg\ell \end{bmatrix}\begin{Bmatrix} x \\ \theta \end{Bmatrix} = \begin{Bmatrix} F \sin \omega t \\ 0 \end{Bmatrix} \tag{7.177}$$

which is identical to Equation 7.172.

7.10 Normal Modes Example

The long but relatively simple example presented in this section, while instructive, does not do justice to the utility of the Method of Normal Modes. Take for example, the problem of determining the response of a system, for which you have a finite element model with several thousands of degrees of freedom, to a specified time-varying force. A specific example is the response of a ship hull to slamming forces caused by the impact of the hull on the water after going over a wave. The model will have thousands of natural frequencies because of the number of degrees of freedom and most of these will not respond to the impact forces which are modeled as forces that decrease exponentially with time as they travel along the length of the hull. The eigenvalues and eigenvectors are readily available from the finite element model and you can choose to solve for only a few of the decoupled modes, thereby saving large amounts of computer time without sacrificing accuracy.

The cart and pendulum example just derived in Section 7.9 is used here to demonstrate the application of the Method of Normal Modes. It is a 2DOF system (x and θ are the degrees of freedom) with a force, $f(t)$, applied to the rolling mass. The linearized equations of motion about the equilibrium state, $x = 0$ and $\theta = 0$, taken from Equation 7.177, are

$$\begin{bmatrix} 2m & m\ell \\ m\ell & m\ell^2 \end{bmatrix}\begin{Bmatrix} \ddot{x} \\ \ddot{\theta} \end{Bmatrix} + \begin{bmatrix} k & 0 \\ 0 & mg\ell \end{bmatrix}\begin{Bmatrix} x \\ \theta \end{Bmatrix} = \begin{Bmatrix} F \sin \omega t \\ 0 \end{Bmatrix}$$

Substituting numbers ($m = 1000$ kg, $k = 100,000$ N/m, $\ell = 0.25$ m, $g = 9.81$ m/s^2, and $f(t) = 1000 \cos(10t)$ N) gives

$$\begin{bmatrix} 2000 & 250 \\ 250 & 62.5 \end{bmatrix} \begin{Bmatrix} \ddot{x} \\ \ddot{\theta} \end{Bmatrix} + \begin{bmatrix} 100000 & 0 \\ 0 & 2452.5 \end{bmatrix} \begin{Bmatrix} x \\ \theta \end{Bmatrix} = \begin{Bmatrix} 1000 \cos(10t) \\ 0 \end{Bmatrix}$$

where the inertia coupling is clear. The goal is to find the solution to this coupled set of differential equations.

Solving the eigenvalue problem gives[10]

$$\text{first mode: } \omega_1^2 = 25.680755 \ (\text{rad/s})^2 \ \{X\}_1 = \begin{Bmatrix} X \\ \Theta \end{Bmatrix}_1 = \begin{Bmatrix} 0.131998 \\ 1.000000 \end{Bmatrix}$$

and

$$\text{second mode: } \omega_2^2 = 152.799245 \ (\text{rad/s})^2 \ \{X\}_2 = \begin{Bmatrix} X \\ \Theta \end{Bmatrix}_2 = \begin{Bmatrix} -0.185798 \\ 1.000000 \end{Bmatrix}$$

The first mode shape is normalized as follows

a. Calculate

$$\{X\}_1^T [M] \{X\}_1 = \{ 0.131998 \ \ 1.000000 \} \begin{bmatrix} 2000 & 250 \\ 250 & 62.5 \end{bmatrix} \begin{Bmatrix} 0.131998 \\ 1.000000 \end{Bmatrix} = 163.346047$$

b. Scale the eigenvector by dividing each element by the square root of the number in part(a)

$$\{X\}_1 = \begin{Bmatrix} X \\ \Theta \end{Bmatrix}_1 = \begin{Bmatrix} 0.131998/\sqrt{163.346047} \\ 1.000000/\sqrt{163.346047} \end{Bmatrix} = \begin{Bmatrix} 0.010328 \\ 0.078243 \end{Bmatrix}$$

c. Now,

$$\{X\}_1^T [M] \{X\}_1 = \{ 0.010328 \ \ 0.078242 \} \begin{bmatrix} 2000 & 250 \\ 250 & 62.5 \end{bmatrix} \begin{Bmatrix} 0.010328 \\ 0.078242 \end{Bmatrix} = 1.000000$$

The second mode shape yields

$$\{X\}_2^T [M] \{X\}_2 = \{ -0.185798 \ \ 1.000000 \} \begin{bmatrix} 2000 & 250 \\ 250 & 62.5 \end{bmatrix} \begin{Bmatrix} -0.185798 \\ 1.000000 \end{Bmatrix}$$

$$= 38.642821$$

and the scaled mode is

$$\{X\}_2 = \begin{Bmatrix} Y \\ \Theta \end{Bmatrix}_2 = \begin{Bmatrix} -0.185798/\sqrt{38.642821} \\ 1.000000/\sqrt{38.642821} \end{Bmatrix} = \begin{Bmatrix} -0.029889 \\ 0.160866 \end{Bmatrix}$$

---

10 I ask the reader to forgive the excessive number of significant figures presented here. That number can, in no way, be justified given the number of significant figures in the input data. Nevertheless, some readers will try to check these numbers with their calculators and I don't want them to be discouraged by round-off errors.

The modal matrix is then

$$[U] = [\{X\}_1 \, \{X\}_2] = \begin{bmatrix} 0.010328 & -0.029889 \\ 0.078242 & 0.160866 \end{bmatrix}$$

The vector of degrees-of-freedom can be expanded in terms of the eigenvectors as

$$\begin{Bmatrix} x(t) \\ \theta(t) \end{Bmatrix} = c_1(t) \begin{Bmatrix} X \\ \Theta \end{Bmatrix}_1 + c_2(t) \begin{Bmatrix} X \\ \Theta \end{Bmatrix}_2$$

where the $c_i(t)$ variables represent the fraction of each mode contributing to the values of the degrees-of-freedom, $x(t)$ and $\theta(t)$, at any time t.

This can also be expressed as

$$\begin{Bmatrix} x(t) \\ \theta(t) \end{Bmatrix} = [U] \begin{Bmatrix} c_1(t) \\ c_2(t) \end{Bmatrix}$$

Substituting this into the equations-of-motion yields

$$[M][U]\{\ddot{c}\} + [K][U]\{c\} = \{f\}$$

Premultiplying this by the transpose of the modal matrix gives

$$[U]^T [M][U]\{\ddot{c}\} + [U]^T [K][U]\{c\} = [U]^T \{f\}$$

where the normalization of the eigenvectors causes

$$[U]^T [M][U] = [I]$$

where $[I]$ is the identity matrix and

$$[U]^T [K][U] = [D(\omega)]$$

where $[D(\omega)]$ is a diagonal matrix with the squared natural frequencies as elements.

For the particular example here

$$[U]^T [M][U] = \begin{bmatrix} 0.010328 & 0.078242 \\ -0.029889 & 0.160866 \end{bmatrix} \begin{bmatrix} 2000 & 250 \\ 250 & 62.5 \end{bmatrix} \begin{bmatrix} 0.010328 & -0.029889 \\ 0.078242 & 0.160866 \end{bmatrix}$$

resulting in

$$[U]^T [M][U] = \begin{bmatrix} 1.000 & 0.000 \\ 0.000 & 1.000 \end{bmatrix}$$

$$[U]^T [K][U] = \begin{bmatrix} 0.010328 & 0.078242 \\ -0.029889 & 0.160866 \end{bmatrix} \begin{bmatrix} 100000 & 0 \\ 0 & 2452.5 \end{bmatrix} \begin{bmatrix} 0.010328 & -0.029889 \\ 0.078242 & 0.160866 \end{bmatrix}$$

giving

$$[U]^T [K][U] = \begin{bmatrix} 25.6808 & 0.00 \\ 0.00 & 152.799 \end{bmatrix}$$

and

$$[U]^T \{f\} = \begin{bmatrix} 0.010328 & 0.078242 \\ -0.029889 & 0.160866 \end{bmatrix} \begin{Bmatrix} 1000\cos(10t) \\ 0 \end{Bmatrix} = \begin{Bmatrix} 10.328\cos(10t) \\ -29.889\cos(10t) \end{Bmatrix}$$

The equations-of-motion then become

$$\begin{bmatrix} 1.00 & 0.00 \\ 0.00 & 1.00 \end{bmatrix} \begin{Bmatrix} \ddot{c}_1 \\ \ddot{c}_2 \end{Bmatrix} + \begin{bmatrix} 25.6808 & 0.00 \\ 0.00 & 152.799 \end{bmatrix} \begin{Bmatrix} c_1 \\ c_2 \end{Bmatrix} = \begin{Bmatrix} 10.328\cos(10t) \\ -29.889\cos(10t) \end{Bmatrix}$$

The system of equations has now been decoupled and can be written as two, independent equations that can be solved, given the initial conditions, using the methods we applied to single degree-of-freedom systems. That is

$$\ddot{c}_1 + 25.6808 c_1 = 10.328\cos(10t)$$

and

$$\ddot{c}_2 + 152.799 c_2 = -29.889\cos(10t)$$

If we let the initial conditions for x and θ be

$$\begin{Bmatrix} x(0) \\ \theta(0) \end{Bmatrix} = \begin{Bmatrix} x_o \\ \theta_o \end{Bmatrix}$$

and

$$\begin{Bmatrix} \dot{x}(0) \\ \dot{\theta}(0) \end{Bmatrix} = \begin{Bmatrix} v_o \\ \omega_o \end{Bmatrix}$$

then the initial conditions on c can be found from Equation 7.142 to be

$$\{c(0)\} = [U]^T [M] \{x(0)\}$$

which gives

$$\begin{Bmatrix} c_1(0) \\ c_2(0) \end{Bmatrix} = \begin{bmatrix} 0.010328 & 0.078242 \\ -0.029889 & 0.160866 \end{bmatrix} \begin{bmatrix} 2000 & 250 \\ 250 & 62.5 \end{bmatrix} \begin{Bmatrix} x_o \\ \theta_o \end{Bmatrix}$$

or

$$\begin{Bmatrix} c_1(0) \\ c_2(0) \end{Bmatrix} = \begin{bmatrix} 40.2165 & 7.4721 \\ -19.5615 & 2.5819 \end{bmatrix} \begin{Bmatrix} x_o \\ \theta_o \end{Bmatrix}$$

Similarly, for the initial velocities,

$$\begin{Bmatrix} \dot{c}_1(0) \\ \dot{c}_2(0) \end{Bmatrix} = \begin{bmatrix} 40.2165 & 7.4721 \\ -19.5615 & 2.5819 \end{bmatrix} \begin{Bmatrix} v_o \\ \omega_o \end{Bmatrix}$$

Given that a solution has been generated out to a time, $t = t_s$, the physical degrees-of-freedom can be recovered using

$$\begin{Bmatrix} x(t_s) \\ \theta(t_s) \end{Bmatrix} = [U] \begin{Bmatrix} c_1(t_s) \\ c_2(t_s) \end{Bmatrix} = \begin{bmatrix} 0.010328 & -0.029889 \\ 0.078242 & 0.160866 \end{bmatrix} \begin{Bmatrix} c_1(t_s) \\ c_2(t_s) \end{Bmatrix}$$

Exercises

7.1 The figure shows two systems, a one degree of freedom system and a two degree of freedom system, with equal masses and equal stiffnesses everywhere. If you write the equations of motion for the systems and solve for the natural frequencies, you will find that $\sqrt{k/m}$ is a natural frequency for both of them. Explain this result by referring to the mode shape of the two degree of freedom system at this frequency.

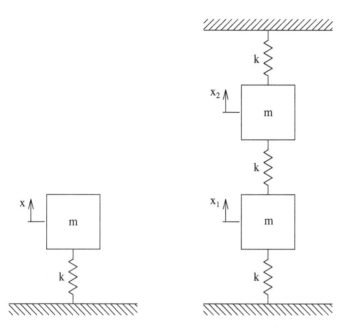

Figure E7.1

7.2 For the system shown in the figure:
   (a) Use Lagrange's Equations to find the equations-of-motion using x_1 and x_2 as the degrees-of-freedom. x_1 and x_2 are small motions away from static equilibrium.
   (b) Use Newton's laws to find the equations-of-motion again and confirm that the two results agree.
   (c) Find the natural frequencies and mode shapes for the system.

7.3 The goal of this exercise is to solidify our understanding of mode shapes and how they are affected by the choice of degrees of freedom. The figure shows two carts of mass, m, and, $2m$, respectively. They are connected to each other by two springs, each of stiffness, k. The larger cart is also connected to the ground by a spring of stiffness, k. The system will be analyzed twice – once using absolute coordinates and then using relative coordinates.
   (a) Use absolute coordinates, x_1 and x_2, as the degrees-of-freedom where x_1 measures the displacement of the larger, supporting cart and x_2 measures the displacement of the small cart away from their static equilibrium positions. Derive the equations of motion and find the two natural frequencies and their corresponding mode shapes.
   (b) Repeat part (a), defining x_1 as an absolute coordinate again (note that there needs to be at least one absolute coordinate for any set of equations of motion) but

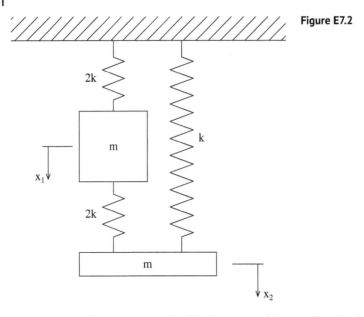

Figure E7.2

now let x_2 measure the displacement of the small cart relative to the larger cart. Find the two natural frequencies and their corresponding mode shapes for these coordinates.

(c) The physical system being analyzed did not change because you used different coordinates to describe it. Do your results confirm that?

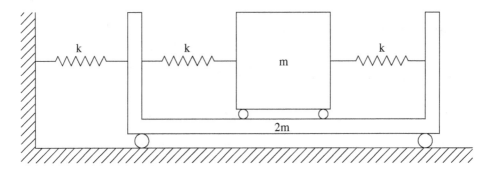

Figure E7.3

7.4 A system with two degrees of freedom has the characteristic equation

$$(\lambda^2 + 6\lambda + 10)(\lambda^2 + 2\lambda + 17) = 0$$

Calculate the undamped natural frequency, damped natural frequency, and damping ratio for each of the two modes.

7.5 The force $F(t) = F \sin \omega t$ is applied as shown in the figure. At certain frequencies, the two masses that do not have forces applied to them will absorb the vibrations and the mass with the force applied to it will stop moving. Write expressions for the two absorbing frequencies in terms of k and m.

Figure E7.5

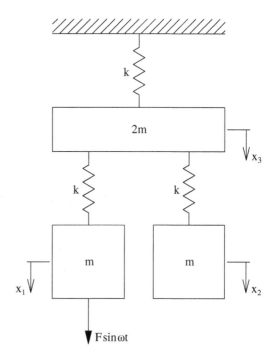

7.6 The simple pendulum shown on the left is subjected to a horizontal harmonic force, $F \sin \omega t$, that causes large amplitude angular motion because the forcing frequency, ω, is identical to the natural frequency of the pendulum. That is, $\omega = \sqrt{g/\ell}$.
A modification to the system, seen on the right, is made by adding another, larger, pendulum and connecting it to the original pendulum via a spring of stiffness, k. The second pendulum is meant to act as a vibration absorber.

(a) Using θ_1 and θ_2 as counterclockwise angular degrees of freedom for the original pendulum and the added pendulum, respectively, show that the equations of motion for the system are

$$\begin{bmatrix} m\ell^2 & 0 \\ 0 & 4M\ell^2 \end{bmatrix} \begin{Bmatrix} \ddot{\theta}_1 \\ \ddot{\theta}_2 \end{Bmatrix} + \begin{bmatrix} mg\ell + k\ell^2 & -2k\ell^2 \\ -2k\ell^2 & 2Mg\ell + 4k\ell^2 \end{bmatrix} \begin{Bmatrix} \theta_1 \\ \theta_2 \end{Bmatrix}$$
$$= \begin{Bmatrix} F\ell \sin \omega t \\ 0 \end{Bmatrix}$$

(b) What value of M, expressed in terms of m, g, k, and ℓ, will make the original pendulum have zero response to the applied force?

7.7 The two carts (masses m and $4m$) are connected together and to the ground by springs of stiffness, k, as shown. There is a force, $f(t)$, applied to the lower cart. Proceed as

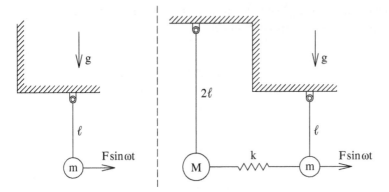

Figure E7.6

follows in order to be able to predict the time response of the carts to the applied force.

(a) Using x_1 and x_2, small motions away from equilibrium, as degrees of freedom, show that the equations of motion for the system are

$$\begin{bmatrix} m & 0 \\ 0 & 4m \end{bmatrix} \begin{Bmatrix} \ddot{x}_1 \\ \ddot{x}_2 \end{Bmatrix} + \begin{bmatrix} 2k & -k \\ -k & k \end{bmatrix} \begin{Bmatrix} x_1 \\ x_2 \end{Bmatrix} = \begin{Bmatrix} 0 \\ f(t) \end{Bmatrix}$$

(b) Find expressions for the natural frequencies and the corresponding mode shapes.
(c) Normalize the mode shapes with respect to the mass matrix and write the modal matrix.
(d) Write the decoupled, differential, equations of motion.
(e) Assuming that you were given a forcing function, $f(t)$, and then solved the decoupled equations of motion, how would you transform back to the physical coordinates?

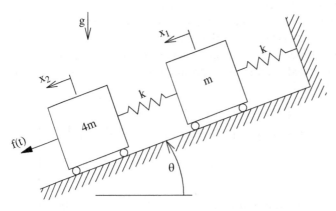

Figure E7.7

8

Continuous Systems

Up to this point, we have had rigid bodies with mass but no elasticity and elastic bodies with no mass. We now consider systems where the mass and stiffness cannot be distinguished from each other. These are called *continuous* or *distributed parameter* systems.

8.1   The Equations of Motion for a Taut String

The first continuous system that we will consider is a string under tension that experiences transverse vibrations. Figure 8.1 shows the string. It has length, L, and is held under constant tension by the force, T. The string has a mass per unit length of m so that its total mass is mL. The transverse displacement of the string is shown by the solid line in the figure. We denote the displacement by $y(x, t)$ to indicate that the displacement of any point on the vibrating string depends on where it is located, x, and what time it is, t. The positive senses of both x and y are shown in the figure and the origin is at the point where the string is attached to the wall.

Consider an infinitesimal element of the string as shown in the inset in Figure 8.1. The element is displaced from its equilibrium position by $y(x, t)$ and has an acceleration $\partial^2 y / \partial t^2$ in the direction shown.

A vertical[1] force balance on the element yields

$$+\uparrow \sum F : -T \sin \theta_1 + T \sin \theta_2 = (mdx)\frac{\partial^2 y}{\partial t^2} \tag{8.1}$$

where (mdx) is the mass of the element, obtained by multiplying the mass per unit length, m, by the length of the element, dx. Since we are considering small motion about equilibrium, we can linearize this by using the small angle approximation, $\sin \theta \approx \theta$, and write

$$-T\theta_1 + T\theta_2 = (mdx)\frac{\partial^2 y}{\partial t^2} \tag{8.2}$$

---

1 You might wonder why the tension is the same on both ends of the element. If you do a horizontal force balance, you will see that the horizontal components of the tension are $T_{left} \cos \theta_1$ to the left and $T_{right} \cos \theta_2$ to the right. Since there is no horizontal acceleration, these must balance each other and, for the small angles we are considering, $\cos \theta_1 = \cos \theta_2 = 1$ so that $T_{left} = T_{right} = T$.

Introduction to Mechanical Vibrations, First Edition. Ronald J. Anderson.
© 2020 John Wiley & Sons Ltd. Published 2020 by John Wiley & Sons Ltd.
Companion website: www.wiley.com/go/anderson/introduction-to-vibrations

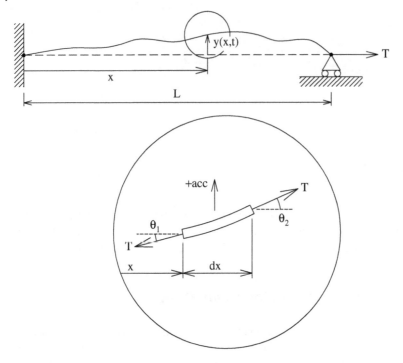

Figure 8.1   A taut string.

The angles, θ_1 and θ_2, are the slopes of the ends of the string. We can express the slope at the left end as

$$\theta_1 = \frac{\partial y}{\partial x} \tag{8.3}$$

and the slope at the right-hand end will be what it is on the left plus the amount that it changes over the length of the element, or

$$\theta_2 = \theta_1 + \frac{\partial \theta}{\partial x}dx = \frac{\partial y}{\partial x} + \frac{\partial^2 y}{\partial x^2}dx \tag{8.4}$$

Substituting Equations 8.3 and 8.4 into Equation 8.2 gives

$$-T\frac{\partial y}{\partial x} + T\left(\frac{\partial y}{\partial x} + \frac{\partial^2 y}{\partial x^2}dx\right) = m\frac{\partial^2 y}{\partial t^2}dx \tag{8.5}$$

which simplifies to

$$T\frac{\partial^2 y}{\partial x^2} = m\frac{\partial^2 y}{\partial t^2} \tag{8.6}$$

Equation 8.6 is the equation of motion for the string and we could work with it directly. However, it is usual at this point to modify it by dividing throughout by m and defining a new parameter, $c^2 = T/m$, giving the equation of motion as

$$c^2\frac{\partial^2 y}{\partial x^2} = \frac{\partial^2 y}{\partial t^2} \tag{8.7}$$

which is variously known as the *1D Wave Equation* or the *String Equation*.

Everything we have done to this point required that we specify initial conditions before we could solve the equations of motion. In the case of continuous systems, we need to specify both initial conditions (displacement and velocity at time zero for every location on the string) and boundary conditions (the displacement at the ends of the string). In general, we will derive governing equations that have partial derivatives up to order 2N (2N = 2 in this case) and we need to specify N boundary conditions at each boundary.

8.2 Natural Frequencies and Mode Shapes for a Taut String

Without knowing the initial conditions, we can ask the question: *Is it possible for every point on the string to have harmonic motion at the same frequency but with different amplitudes?* This is precisely the question we asked for NDOF systems and the answer was "yes". It is here as well.

Mathematically, what we are asking is, can

$$y(x, t) = Y(x) \cos \omega t \tag{8.8}$$

be a solution? What we have done is define a continuous function of x, $Y(x)$, that will give us the shape of the string if we can find a frequency, ω.

The time derivative we need in Equation 8.7, is

$$\frac{\partial^2 y}{\partial t^2} = -\omega^2 Y(x) \cos \omega t \tag{8.9}$$

and the position derivative is

$$\frac{\partial^2 y}{\partial x^2} = \frac{d^2 Y(x)}{dx^2} \cos \omega t \tag{8.10}$$

Substituting these expressions into Equation 8.7 gives

$$c^2 \frac{d^2 Y(x)}{dx^2} \cos \omega t = -\omega^2 Y(x) \cos \omega t \tag{8.11}$$

We can cancel the cos ωt terms, divide by c^2, and, simply for convenience, use Y to represent $Y(x)$, resulting in the following second-order, ordinary, differential equation

$$\frac{d^2 Y}{dx^2} + \left(\frac{\omega}{c}\right)^2 Y = 0. \tag{8.12}$$

with boundary conditions $Y(0) = 0$ and $Y(L) = 0$

The solution to Equation 8.12 needs to be a shape, $Y(x)$, that starts at zero when $x = 0$ and then returns to zero when $x = L$ in order to satisfy the boundary conditions. In addition to satisfying the boundary conditions, $Y(x)$ must repeat itself, with a sign change, on its second derivative in order to make the two terms in Equation 8.12 add up to zero. There are two functions that repeat themselves with a sign change on the second derivative – the cos kx function and the sin kx function. Of these, only the sin kx function is zero at $x = 0$ so our candidate for the shape is

$$Y(x) = A \sin kx \tag{8.13}$$

where A and k are to be determined.

We differentiate twice to get

$$\frac{d^2Y}{dx^2} = -k^2A \ \sin \ kx \tag{8.14}$$

and substitute into Equation 8.12, yielding

$$-k^2A \ \sin \ kx + \left(\frac{\omega}{c}\right)^2 A \ \sin \ kx = 0 \tag{8.15}$$

or

$$\left[\left(\frac{\omega}{c}\right)^2 - k^2\right] A \ \sin \ kx = 0 \tag{8.16}$$

Equation 8.12 can be satisfied by $A = 0$ but that would mean zero amplitude everywhere so we dismiss it as a solution. $\sin \ kx$ is not zero everywhere so we can cancel it out. That leaves us with

$$k = \frac{\omega}{c} \tag{8.17}$$

and the solution is therefore of the form

$$Y(x) = A \ \sin \ \left(\frac{\omega x}{c}\right) \tag{8.18}$$

where it remains to satisfy the boundary condition at $x = L$. Setting $x = L$ and $Y(L) = 0$ in Equation 8.18 gives the condition

$$A \ \sin \ \left(\frac{\omega L}{c}\right) = 0 \tag{8.19}$$

We again dismiss $A = 0$ as an option. The boundary condition is satisfied for

$$\sin \ \left(\frac{\omega L}{c}\right) = 0 \tag{8.20}$$

and this requires that

$$\frac{\omega L}{c} = n\pi \quad ; \ n = 0, 1, 2, \ldots \tag{8.21}$$

The result of all this is that the answer to our original question "Is it possible for every point on the string to have harmonic motion at the same frequency but with different amplitudes?" is *Yes but only for certain frequencies.* The frequencies are the natural frequencies of the string and there are infinitely many of them. The natural frequencies are

$$\omega_n = \frac{n\pi c}{L} \quad ; \ n = 0, 1, 2, \ldots \tag{8.22}$$

We defined $c^2 = T/m$ just before Equation 8.7 so we can substitute the original parameters back now to get

$$\omega_n = \frac{n\pi}{L}\sqrt{\frac{T}{m}} \quad ; \ n = 0, 1, 2, \ldots \tag{8.23}$$

Now that we have the natural frequencies we can attempt to find the mode shapes. We said in Equation 8.18 that the shape was

$$Y(x) = A \ \sin \ \left(\frac{\omega x}{c}\right) \tag{8.24}$$

We can now substitute the natural frequencies from Equation 8.22 to say that the mode shapes are

$$Y_n(x) = A \sin\left(\frac{n\pi x}{L}\right) \quad ; n = 0, 1, 2, \ldots \tag{8.25}$$

where we recognize that we will never be able to find the amplitude, A, because we didn't specify any initial conditions. We can choose arbitrary values for A and scale the mode shapes as we did for NDOF systems. We typically choose $A = 1$ and write the mode shapes as

$$Y_n(x) = \sin\left(\frac{n\pi x}{L}\right) \quad ; n = 0, 1, 2, \ldots \tag{8.26}$$

Figure 8.2 shows the first five modes for the string. The string is shown at its extreme positive and negative displacements. The points with zero displacement are called *nodal points* or, simply, *nodes* and they are points that never see any motion. The string between the nodes oscillates back and forth at the natural frequencies. Notice that higher-frequency modes have more nodes than lower-frequency modes. The mode for $n = 0$ is not shown because it has zero frequency and the mode shape is a horizontal line indicating that the string being in equilibrium is a solution, as it always must be.

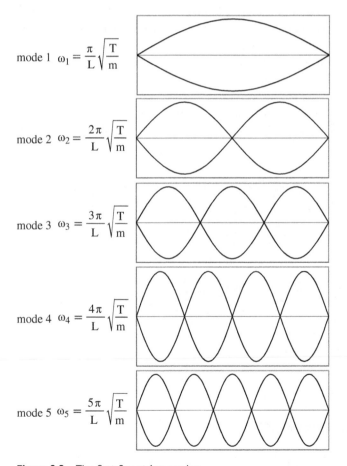

mode 1 $\omega_1 = \dfrac{\pi}{L}\sqrt{\dfrac{T}{m}}$

mode 2 $\omega_2 = \dfrac{2\pi}{L}\sqrt{\dfrac{T}{m}}$

mode 3 $\omega_3 = \dfrac{3\pi}{L}\sqrt{\dfrac{T}{m}}$

mode 4 $\omega_4 = \dfrac{4\pi}{L}\sqrt{\dfrac{T}{m}}$

mode 5 $\omega_5 = \dfrac{5\pi}{L}\sqrt{\dfrac{T}{m}}$

Figure 8.2 The first five string modes.

8.3 Vibrations of Uniform Beams

We move on to consider another continuous system – the uniform beam[2]. The beam is shown in Figure 8.3 where the beam parameters are: length, L, Young's modulus, E, area moment of inertia, I, and mass per unit length, m.

Figure 8.3 also defines the degrees of freedom for the beam. There are two independent variables, the position, x, which locates sections of the beam and time, t. The transverse deflection of a section is denoted by $y(x, t)$. For a section located a distance x from the left end of the beam, $y(x, t)$ is a measure of the instantaneous distance, at time t, from the equilibrium position of the neutral axis to its deflected position. The positive directions for x and y are to the right and downward respectively. Also shown on Figure 8.3 is the local slope, $\theta(x, t)$, which is defined mathematically by

$$\theta(x, t) = \frac{\partial y(x, t)}{\partial x} \tag{8.27}$$

Finally, Figure 8.3 shows the boundary conditions on the beam. There are two types of boundary conditions. The first type is related to geometrical variables and these are shown on the deflected beam in the center of the figure. They are: $y(0, t)$ and $y(L, t)$ which specify the displacements at the left and right ends of the beam respectively and $\theta(0, t)$ and $\theta(L, t)$ which specify the slopes at the ends. The second type of boundary condition is related to the forces and moments acting at the ends of the beam. These are shown on the lowest beam and are: the bending moments at the ends of the beam, $M(0, t)$ and $M(L, t)$, and the shear forces at the ends, $V(0, t)$ and $V(L, t)$.

Figure 8.4 shows an element of the beam at a general section, x. This is an infinitesimally small element of length, dx. The right side of the figure shows a free body diagram of the

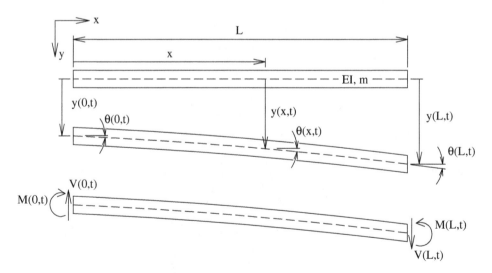

Figure 8.3 Deflections of a uniform beam.

---

2 The beam is uniform in that the *flexural rigidity*, EI, is constant along its length. Non-uniform beams can only be handled by numerical modeling techniques such as the finite element method.

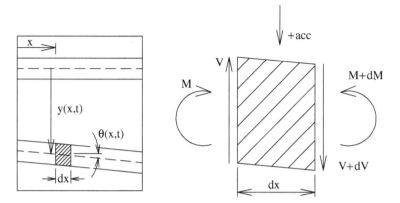

Figure 8.4 A beam element.

element. Since the element has been removed from the beam, the internal shear forces and bending moments are shown on the free body diagram. On the left are the bending moment, M, and shear force, V, at section x. On the right, we see the bending moment and shear force at $x + dx$ which are represented by their values on the left plus an infinitesimally small change that occurs across the element. They are represented by $M + dM$ and $V + dV$. Also shown on the free body diagram is the positive sense of the acceleration of the element.

Summing forces in the downward direction so as to agree with the positive acceleration direction gives

$$-V + (V + dV) = dm\ a \tag{8.28}$$

where dm is the mass of the element and a is its acceleration. We can replace dm with mdx, the mass per unit length multiplied by the length of the element, dV with $(\partial V/\partial x)dx$, and the acceleration with

$$a = \frac{\partial^2 y}{\partial t^2} \tag{8.29}$$

giving

$$-V + \left(V + \frac{\partial V}{\partial x}dx\right) = m\frac{\partial^2 y}{\partial t^2}dx \tag{8.30}$$

which simplifies to

$$\frac{\partial V}{\partial x} = m\frac{\partial^2 y}{\partial t^2} \tag{8.31}$$

Summing moments[3] about the right-hand face of the element in a clockwise direction gives

$$M + Vdx - (M + dM) = 0 \tag{8.32}$$

where we can substitute $(\partial M/\partial x)dx$ for dM and simplify to

$$V - \frac{\partial M}{\partial x} = 0 \tag{8.33}$$

---

3 We use a static moment balance here under the assumption that the moment of inertia of the element is very small. This is a good approximation for slender beams.

Equation 8.33 can be differentiated with respect to x to give

$$\frac{\partial V}{\partial x} = \frac{\partial^2 M}{\partial x^2} \tag{8.34}$$

and this can be substituted into Equation 8.31 to give

$$\frac{\partial^2 M}{\partial x^2} = m\frac{\partial^2 y}{\partial t^2} \tag{8.35}$$

Now we need to relate the bending moment to the displacement and we do that by introducing the following derived constitutive relationship[4] that can be found in textbooks on solid mechanics

$$M = -EI\frac{\partial^2 y}{\partial x^2} \tag{8.36}$$

Differentiating Equation 8.36 twice gives

$$\frac{\partial^2 M}{\partial x^2} = -\frac{\partial^2}{\partial x^2}\left(EI\frac{\partial^2 y}{\partial x^2}\right) \tag{8.37}$$

and it is here that we need to restrict ourselves to uniform beams where EI is constant so that we can bring it outside the differentiation and write Equation 8.37 as

$$\frac{\partial^2 M}{\partial x^2} = -EI\frac{\partial^4 y}{\partial x^4} \tag{8.38}$$

Equation 8.38 can be substituted into Equation 8.35 to yield

$$-EI\frac{\partial^4 y}{\partial x^4} = m\frac{\partial^2 y}{\partial t^2} \tag{8.39}$$

Once again, we ask the eigenvalue question *Is it possible for every point on the beam to oscillate with the same frequency but different amplitudes?* If it is possible, we should be able to express the deflection as

$$y(x, t) = Y(x)\sin \omega t \tag{8.40}$$

Then, differentiating twice with respect to time, yields

$$\frac{\partial^2 y}{\partial t^2} = -\omega^2 Y(x)\sin \omega t \tag{8.41}$$

which can be substituted into Equation 8.39 to give

$$-EI\frac{d^4 Y(x)}{dx^4}\sin \omega t = -m\omega^2 Y(x)\sin \omega t \tag{8.42}$$

---

4 The derivation starts with the stress/strain relationship

$$\sigma = E\epsilon$$

where σ is the stress, ϵ is the strain, and E is Young's modulus. It proceeds to a definition of strain based on curvature that requires definitions of positive directions for x, y, and the bending moment. The negative sign seen here comes from the directions we assumed for positive x, y and M. If we had taken y to be positive up, for instance, we would need to use

$$M = +EI\frac{\partial^2 y}{\partial x^2}$$

This can be simplified to

$$\frac{d^4Y(x)}{dx^4} - \frac{m\omega^2}{EI}Y(x) = 0 \tag{8.43}$$

At this point we define the group of parameters in Equation 8.43 as

$$n^4 = \frac{m\omega^2}{EI} \tag{8.44}$$

so that we can consider the fourth-order, ordinary, differential equation

$$\frac{d^4Y(x)}{dx^4} - n^4Y(x) = 0 \tag{8.45}$$

As we did for all previous differential equations, we consider what possible functions could satisfy Equation 8.45, and arrive at the conclusion that the exponential function will repeat itself on the fourth derivative and is therefore a candidate for adding a function to its fourth derivative and getting zero as the result. Therefore, we assume

$$Y(x) = Ae^{ax} \tag{8.46}$$

in which case, its fourth derivative is

$$\frac{d^4Y(x)}{dx^4} = a^4Ae^{ax} \tag{8.47}$$

and, after substituting Equations 8.46 and 8.47 into Equation 8.45, we get

$$(a^4 - n^4)Ae^{ax} = 0 \tag{8.48}$$

Once again we recognize that $e^{ax} \neq 0$ and $A = 0$ gives us no amplitude so that solutions will only exist when

$$a^4 - n^4 = 0 \tag{8.49}$$

$a^4 - n^4$ is equal to zero for $a = \pm n$ and for $a = \pm in$ where $i = \sqrt{-1}$. As a result, the solutions will be of the forms[5]

$$Y(x) = Ae^{nx} = A(\cosh\ nx + \sinh\ nx) \tag{8.50}$$

$$Y(x) = Ae^{-nx} = A(\cosh\ nx - \sinh\ nx) \tag{8.51}$$

$$Y(x) = Ae^{inx} = A(\cos\ nx + i\ \sin\ nx) \tag{8.52}$$

$$Y(x) = Ae^{-inx} = A(\cos\ nx - i\ \sin\ nx) \tag{8.53}$$

---

5 The definitions of the hyperbolic sine and cosine functions are

$$\sinh\ nx = \frac{e^{nx} - e^{-nx}}{2}$$

and

$$\cosh\ nx = \frac{e^{nx} + e^{-nx}}{2}$$

The identities used in Equations 8.50 and 8.51 can easily be derived from these definitions.

For the linear, differential equation we are considering, the total solution will be formed from the superposition of all the solutions. We therefore write

$$Y(x) = A \cosh nx + B \sinh nx + C \cos nx + D \sin nx \tag{8.54}$$

where A, B, C, D, and n are to be found from the boundary conditions.

Since the governing differential equation (Equation 8.45) is of fourth order we will need two boundary conditions at each end of the beam (see the last paragraph in Section 8.1). The boundary conditions are specifications of displacement, slope, bending moment or shear force at the ends of the beam. The natural frequencies and mode shapes that we find will depend on the boundary conditions so that the same beam with different end constraints will have different natural frequencies. For illustrative purposes, we will consider here the case of a cantilever beam as shown in Figure 8.5 to illustrate the method and the results.

Physically, the boundary conditions are that the beam must have zero deflection and zero slope where it is attached to the wall. Clearly there will be a bending moment and shear force supplied by the wall in order to enforce this constraint but they will be unknown. At the free end of the beam there will be zero bending moment and zero shear force because there is nothing there to apply forces and moments. The displacement and slope at the free end will be non-zero.

The boundary conditions related to the variable we are working with (i.e. displacement) and its first derivative are called *essential* boundary conditions and we can write these immediately as

$$Y(0) = 0 \tag{8.55}$$

and

$$\frac{dY(0)}{dx} = 0 \tag{8.56}$$

The boundary conditions related to the force and moment variables are called *additional* boundary equations and we need to search through the equations we have been using in order to express them in terms of the displacement variable. The bending moment condition can be written using Equation 8.36 as

$$\frac{d^2Y(L)}{dx^2} = 0 \tag{8.57}$$

and Equations 8.33 and 8.36 can be combined to write the shear force condition as

$$\frac{d^3Y(L)}{dx^3} = 0 \tag{8.58}$$

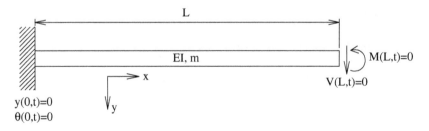

$$L$$

EI, m

$$x$$

$$M(L,t)=0$$

$$V(L,t)=0$$

y(0,t)=0
θ(0,t)=0

y

Figure 8.5 A cantilever beam.

To apply the boundary conditions, we first need to differentiate $Y(x)$ three times. The results are:

$$Y(x) = A \cosh nx + B \sinh nx + C \cos nx + D \sin nx \tag{8.59}$$

$$\frac{dY(x)}{dx} = nA \sinh nx + nB \cosh nx - nC \sin nx + nD \cos nx \tag{8.60}$$

$$\frac{d^2Y(x)}{dx^2} = n^2A \cosh nx + n^2B \sinh nx - n^2C \cos nx - n^2D \sin nx \tag{8.61}$$

$$\frac{d^3Y(x)}{dx^3} = n^3A \sinh nx + n^3B \cosh nx + n^3C \sin nx - n^3D \cos nx \tag{8.62}$$

We can apply the zero deflection boundary condition, $Y(0) = 0$, by substituting $x = 0$ into Equation 8.59, giving

$$Y(0) = A \cosh 0 + B \sinh 0 + C \cos 0 + D \sin 0 = A + C = 0 \tag{8.63}$$

and the zero slope condition is Equation 8.60 with $x = 0$, as follows

$$\frac{dY(0)}{dx} = nA \sinh 0 + nB \cosh 0 - nC \sin 0 + nD \cos 0 = nB + nD = 0 \tag{8.64}$$

Equations 8.63 and 8.64 indicate that $C = -A$ and $D = -B$. We modify Equations 8.61 and 8.62 by replacing C and D with $-A$ and $-B$ respectively before applying the bending moment and shear boundary conditions. The result is

$$\frac{d^2Y(x)}{dx^2} = n^2A(\cosh nx + \cos nx) + n^2B(\sinh nx + \sin nx) \tag{8.65}$$

and

$$\frac{d^3Y(x)}{dx^3} = n^3A(\sinh nx - \sin nx) + n^3B(\cosh nx + \cos nx) \tag{8.66}$$

The bending moment boundary condition (Equation 8.57) becomes

$$\frac{d^2Y(L)}{dx^2} = n^2A(\cosh nL + \cos nL) + n^2B(\sinh nL + \sin nL) = 0 \tag{8.67}$$

or

$$A(\cosh nL + \cos nL) + B(\sinh nL + \sin nL) = 0 \tag{8.68}$$

The shear force boundary condition (Equation 8.58) can be written as

$$\frac{d^3Y(L)}{dx^3} = n^3A(\sinh nL - \sin nL) + n^3B(\cosh nL + \cos nL) = 0 \tag{8.69}$$

or

$$A(\sinh nL - \sin nL) + B(\cosh nL + \cos nL) = 0 \tag{8.70}$$

Equations 8.68 and 8.70 are two equations with three unknowns (A, B, and n) as is typical of eigenvalue problems – there is never enough information to determine everything. The natural frequency is hiding inside n (see Equation 8.44) and we can extract it by writing two expressions for A/B, one from each of Equations 8.68 and 8.70, and equating them. This will leave us with one equation with only n unknown and we should be able to find it.

From Equation 8.68, we get

$$\frac{A}{B} = -\frac{(\sinh\ nL + \sin\ nL)}{(\cosh\ nL + \cos\ nL)} \tag{8.71}$$

and, from Equation 8.70, we get

$$\frac{A}{B} = -\frac{(\cosh\ nL + \cos\ nL)}{(\sinh\ nL - \sin\ nL)} \tag{8.72}$$

Equating the two expressions for A/B gives

$$-\frac{(\sinh\ nL + \sin\ nL)}{(\cosh\ nL + \cos\ nL)} = -\frac{(\cosh\ nL + \cos\ nL)}{(\sinh\ nL - \sin\ nL)} \tag{8.73}$$

Cross multiplying yields

$$(\sinh\ nL + \sin\ nL)(\sinh\ nL - \sin\ nL)$$
$$= (\cosh\ nL + \cos\ nL)(\cosh\ nL + \cos\ nL) \tag{8.74}$$

and this can be expanded to

$$\sinh^2 nL + \sin\ nL\ \sinh\ nL - \sin\ nL\ \sinh\ nL - \sin^2 nL$$
$$= \cosh^2 nL + 2\ \cos\ nL\ \cosh\ nL + \cos^2 nL \tag{8.75}$$

the $\sin\ nL\ \sinh\ nL$ terms cancel each other out and the remainder of the equation can be rearranged as

$$0 = (\cosh^2 nL - \sinh^2 nL) + 2\ \cos\ nL\ \cosh\ nL + (\cos^2 nL + \sin^2 nL) \tag{8.76}$$

We can call on two identities at this point, as follows

$$\cosh^2 nL - \sinh^2 nL = 1 \tag{8.77}$$

and

$$\cos^2 nL + \sin^2 nL = 1 \tag{8.78}$$

so that Equation 8.76 becomes

$$0 = 1 + 2\ \cos\ nL\ \cosh\ nL + 1 \tag{8.79}$$

or, finally,

$$\cos\ nL\ \cosh\ nL = -1 \tag{8.80}$$

Equation 8.80 is the final analytical result. It is a transcendental equation with an infinite number of solutions that can only be found numerically. The first three solutions are

$$(nL)_1 = 1.875 \tag{8.81}$$
$$(nL)_2 = 4.694$$
$$(nL)_3 = 7.855$$

Now that we have values for nL, we can retrieve the definition of n from Equation 8.44, repeated here as Equation 8.82

$$n^4 = \frac{m\omega^2}{EI} \tag{8.82}$$

to see if we can find the natural frequencies. Equation 8.82 can be rearranged to give

$$\omega^2 = n^4 \frac{EI}{m} \tag{8.83}$$

from which we can find ω by taking the square root to get

$$\omega = n^2 \sqrt{\frac{EI}{m}} \tag{8.84}$$

Since we have values for nL rather than for n, we rewrite Equation 8.84 as

$$\omega = \frac{(nL)^2}{L^2} \sqrt{\frac{EI}{m}} \tag{8.85}$$

and, taking all of the system parameters inside the square root sign, we get the final expression for the natural frequencies to be

$$\omega = (nL)^2 \sqrt{\frac{EI}{mL^4}} \tag{8.86}$$

Referring back to the numerical values of nL, we can write the first three natural frequencies as

$$\omega_1 = (1.875)^2 \sqrt{\frac{EI}{mL^4}} \tag{8.87}$$

$$\omega_2 = (4.694)^2 \sqrt{\frac{EI}{mL^4}}$$

$$\omega_3 = (7.855)^2 \sqrt{\frac{EI}{mL^4}}$$

It remains to determine the mode shapes for the cantilever beam. The shape is as specified in Equation 8.54 with the additional information that was derived from the boundary conditions. That is, $C = -A$ and $D = -B$ from Equations 8.63 and 8.64. Using this, Equation 8.59 can be rewritten as

$$Y(x) = A \cosh nx + B \sinh nx - A \cos nx - B \sin nx \tag{8.88}$$

We can then divide throughout by B to get

$$\left(\frac{1}{B}\right) Y(x) = \left(\frac{A}{B}\right) \cosh nx + \sinh nx - \left(\frac{A}{B}\right) \cos nx - \sin nx \tag{8.89}$$

where A/B is known from either of Equations 8.71 or 8.72 providing that we know nL, which we now do.

First rearrange Equation 8.89 as

$$\left(\frac{1}{B}\right) Y(x) = \left(\frac{A}{B}\right)(\cosh nx - \cos nx) + (\sinh nx - \sin nx) \tag{8.90}$$

Then, choosing to use Equation 8.71 for A/B, we can write

$$\left(\frac{1}{B}\right) Y(x) = -\frac{(\sinh nL + \sin nL)}{(\cosh nL + \cos nL)}(\cosh nx - \cos nx) + (\sinh nx - \sin nx) \tag{8.91}$$

and, finally, we recognize that B will never be able to be specified because we don't know the initial conditions.

This has been so for all mode shapes. They are simply relative amplitudes of degrees of freedom and, in this case, we have an infinite number of degrees of freedom so the mode shape is a continuous function. We set $B = 1$ and write the mode shape as

$$Y(x) = -\frac{(\sinh nL + \sin nL)}{(\cosh nL + \cos nL)}(\cosh nx - \cos nx) + (\sinh nx - \sin nx) \qquad (8.92)$$

In order to be able to plot the mode shapes we need to replace the nx terms with $(nL)(x/L)$ because we have numerical values for nL but not for n. The mode shape becomes

$$Y(x) = -\frac{(\sinh nL + \sin nL)}{(\cosh nL + \cos nL)}\left[\cosh\left(nL\frac{x}{L}\right) - \cos\left(nL\frac{x}{L}\right)\right]$$
$$+ \left[\sinh\left(nL\frac{x}{L}\right) - \sin\left(nL\frac{x}{L}\right)\right] \qquad (8.93)$$

Figure 8.6 shows the first three modes for the cantilever beam. The lowest frequency mode is the "wing flapping" mode that one would expect. Higher-frequency modes have nodal points as seen for modes two and three. If even higher-frequency modes were shown you would see that each one has one more nodal point than the mode preceding

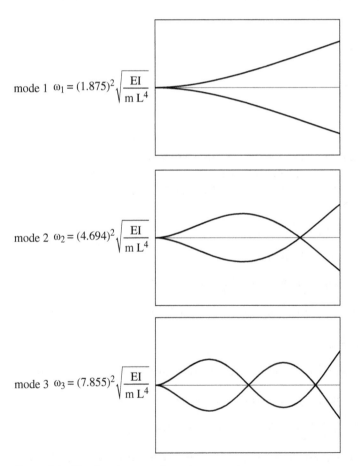

mode 1 $\omega_1 = (1.875)^2\sqrt{\dfrac{EI}{m\,L^4}}$

mode 2 $\omega_2 = (4.694)^2\sqrt{\dfrac{EI}{m\,L^4}}$

mode 3 $\omega_3 = (7.855)^2\sqrt{\dfrac{EI}{m\,L^4}}$

Figure 8.6 The first three cantilever beam modes.

it. Physically, the nodal points get so close together at high frequencies that amplitudes of motion are severely restricted. Practically speaking, we are therefore only interested in the first few modes for any continuous system.

Beams are just a simple example of continuous systems. Every structure has natural frequencies and the natural frequencies have modes. There is an entire field of experimentation called *modal analysis* where engineers determine the natural frequencies and mode shapes of continuous structures such as automotive components experimentally. We have just touched on the theory here.

Exercises

8.1 We considered vibrations of uniform beams in Section 8.3 and used a cantilever beam as an example. Everything done there, up to and including Equation 8.54, is applicable to any uniform beam. Consider now the case of a simply supported, uniform, beam.
  (a) What are the boundary conditions?
  (b) Find the natural frequencies.
  (c) Find the mode shapes.

8.2 A free-free beam is one where both ends are free of any geometrical constraints. That is, the deflection and slope at the boundaries are not specified in any way. The boundary conditions are that the shear force and bending moment are zero at the ends of the beam. One example of a free-free beam is a ship floating in the ocean. Another is a beam supported by an elastic foundation such as a railway rail.
  (a) Derive the transcendental equation for the free-free beam (similar to Equation 8.80 for the cantilever beam).
  (b) Find the first few natural frequencies numerically.

8.3 Calculate the first four natural frequencies for a 1 m long, rectangular, steel, cantilever beam with a base, $b = 20$ mm, and a height, $h = 5$ mm. The properties of steel are: density, $\rho = 7850$ kg/m^3; modulus of elasticity, $E = 200$ GPa. The area moment of inertia of a rectangular section is $I = bh^3/12$.

9

Finite Elements

The finite element method is widely used in engineering analysis today and it is expected that all graduating engineers will be familiar with it. Basically, it is a method for approximating the solution to complex engineering problems by saying that all intractable problems can be broken down into a set of solvable problems in small elements of the overall system and then the solutions in the elements can be brought together to get the solution for the whole.

9.1 Shape Functions

Consider the case of a uniform beam in bending. Without being specific about the geometry of the beam or the loads acting on it, Figure 9.1 shows an infinitesimally small element of the beam. A force balance on the element gives

$$V - \left(V + \frac{dV}{dx} dx \right) = 0 \tag{9.1}$$

or

$$\frac{dV}{dx} = 0 \tag{9.2}$$

A moment balance about the right-hand face of the element gives

$$M + V dx - \left(M + \frac{dM}{dx} dx \right) = 0 \tag{9.3}$$

or

$$V = \frac{dM}{dx} \tag{9.4}$$

We can differentiate Equation 9.4 to get

$$\frac{dV}{dx} = \frac{d^2M}{dx^2} \tag{9.5}$$

and equate this to Equation 9.2, giving

$$\frac{d^2M}{dx^2} = 0 \tag{9.6}$$

Introduction to Mechanical Vibrations, First Edition. Ronald J. Anderson.
© 2020 John Wiley & Sons Ltd. Published 2020 by John Wiley & Sons Ltd.
Companion website: www.wiley.com/go/anderson/introduction-to-vibrations

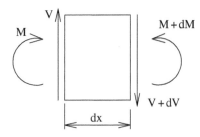

Figure 9.1 A beam element.

We then introduce the constitutive relationship for beam bending

$$M = \pm EI \frac{d^2y}{dx^2} \qquad (9.7)$$

where y is the beam deflection at section x and the $\pm$ indicates that the sign on the constitutive relationship depends on the positive directions chosen for x and y.

Differentiating Equation 9.7 twice, knowing that EI is a constant for the uniform
beam, gives

$$\frac{d^2M}{dx^2} = \pm EI \frac{d^4y}{dx^4} \qquad (9.8)$$

Equation 9.8 and Equation 9.6 can be combined to give

$$\frac{d^4y}{dx^4} = 0 \qquad (9.9)$$

Equation 9.9 can be solved by integration. The first integration yields

$$\frac{d^3y}{dx^3} = c_1 \qquad (9.10)$$

The second results in

$$\frac{d^2y}{dx^2} = c_1x + c_2 \qquad (9.11)$$

The third gives

$$\frac{dy}{dx} = \frac{c_1}{2}x^2 + c_2x + c_3 \qquad (9.12)$$

and, finally,

$$y = \frac{c_1}{6}x^3 + \frac{c_2}{2}x^2 + c_3x + c_4 \qquad (9.13)$$

The point is that the deflection of the uniform beam can be expressed as a cubic function
of the distance from its origin. The cubic function is called the "shape function" for the
beam and we go forward under the premise that beams deflect as cubic functions[1].

Consider now a non-uniform beam with multiple loads and supports such as that shown
in Figure 9.2. The deflection of the beam from $x = 0$ to $x = L$ clearly cannot be approximated
by a single cubic function but, if we divide the beam into short sections, the deflection in

---

1 In the days before drawings were done on computers, engineers had a flexible ruler that could be used to
draw smooth curves passing through points on a graph. These were simply soft beams that deflected as
cubic functions and they have now been replaced in software by "cubic spline" functions.

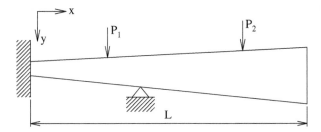

Figure 9.2 A non-uniform beam.

each section will behave approximately as a cubic function. We need to ensure continuity between the short sections so we define finite elements of the beam and connect them at nodes where we look after the continuity requirements.

The assumed deflected shape of the element is the "shape function" that is the basis for finite element analysis.

9.2 The Stiffness Matrix for an Elastic Rod

We start our look at the finite element method with an elastic rod. The reason for using this as a starting example is that it is sufficiently simple to work through analytically while still retaining all of the properties of other, more complex, finite elements.

First we consider a static analysis of the rod of length, L, shown in Figure 9.3 with the force P applied to it with the resulting deflection, δ. Let A be the cross-sectional area of the

Figure 9.3 An elastic rod.

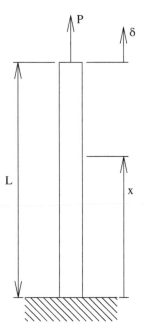

rod and E be Young's modulus. The stress in the rod is then

$$\sigma = \frac{P}{A} \qquad (9.14)$$

and the strain is

$$\epsilon = \frac{\delta}{L} \qquad (9.15)$$

Inserting these into the stress/strain relationship

$$\sigma = E\epsilon \qquad (9.16)$$

gives

$$\frac{P}{A} = E\frac{\delta}{L} \qquad (9.17)$$

or

$$\delta = \frac{PL}{AE} \qquad (9.18)$$

The deflection at any section x of the rod can be expressed as a linear function

$$\delta(x) = \frac{P}{AE}x \qquad (9.19)$$

giving $\delta(0) = 0$ and $\delta(L) = PL/AE$ as expected.

Now consider making the rod into an element with nodes at its ends. Figure 9.4 shows the element with two nodes, numbered 1 and 2, and end deflections, u_1 and u_2.

We know from the static analysis just done that the displacement varies linearly along the element so we can write the linear shape function

$$u(x) = c_1 x + c_2 \qquad (9.20)$$

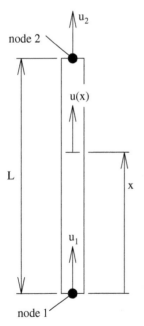

Figure 9.4 An elastic rod element.

node 2

u_2

$u(x)$

L

x

u_1

node 1

The boundary conditions are $u(0) = u_1$ and $u(L) = u_2$, resulting in

$$u(0) = c_2 = u_1 \tag{9.21}$$

and

$$u(L) = c_1 L + c_2 = u_2 \tag{9.22}$$

Equation 9.21 can be substituted into Equation 9.22 to find

$$c_1 = \frac{u_2 - u_1}{L} \tag{9.23}$$

The displacement at any point in the element can then be written using Equation 9.20 with the values of c_1 and c_2 from Equations 9.23 and 9.21 as

$$u(x) = u_1 \left(1 - \frac{x}{L}\right) + u_2 \left(\frac{x}{L}\right) \tag{9.24}$$

We are going to develop the equations of motion from Lagrange's Equation so we can start by looking at the potential energy (also known as strain energy) of the rod element. Figure 9.5 shows an infinitesimal element taken out of the finite element of Figure 9.4. The element has length, dx, and is loaded by the force, P. The change in length of the element due to the applied force is du. The stress is

$$\sigma = \frac{P}{A} \tag{9.25}$$

where A is the cross-sectional area. The strain is the change in length divided by the original length, or

$$\epsilon = \frac{du}{dx} \tag{9.26}$$

and the stress/strain relationship, $\sigma = E\epsilon$, results in

$$\frac{P}{A} = E\frac{du}{dx} \tag{9.27}$$

Figure 9.5 An element of the element.

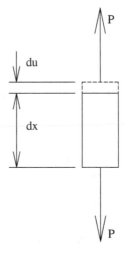

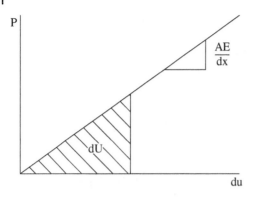

Figure 9.6 *P* versus *du* for the element.

which can be rearranged to

$$P = \left(\frac{AE}{dx}\right) du \tag{9.28}$$

The relationship in Equation 9.28 is plotted in Figure 9.6 and the infinitesimal amount of strain energy, dU, is the shaded triangular area under the curve[2].

We can write the triangular area as one half its base, du, multiplied by its height, $P = (AE/dx)du$, to get

$$dU = \frac{1}{2}(du)\left(\frac{AE}{dx}\right) du \tag{9.29}$$

which can be multiplied and divided by dx and rearranged to get

$$dU = \frac{1}{2}AE\left(\frac{du}{dx}\right)^2 dx \tag{9.30}$$

The expression in Equation 9.30 can be integrated over the finite element to give the total strain energy, U, as

$$U = \int_{x=0}^{x=L} \frac{1}{2}AE\left(\frac{du}{dx}\right)^2 dx = \frac{AE}{2}\int_{x=0}^{x=L} \left(\frac{du}{dx}\right)^2 dx \tag{9.31}$$

Equation 9.24 can be differentiated to give

$$\frac{du}{dx} = -\frac{u_1}{L} + \frac{u_2}{L} \tag{9.32}$$

which can be squared to yield

$$\left(\frac{du}{dx}\right)^2 = \frac{1}{L^2}(u_1^2 - 2u_1 u_2 + u_2^2) \tag{9.33}$$

Substituting Equation 9.33 into Equation 9.31 gives

$$U = \frac{AE}{2L^2}\int_{x=0}^{x=L} (u_1^2 - 2u_1 u_2 + u_2^2)dx \tag{9.34}$$

and the integral can be evaluated to find

$$U = \frac{AE}{2L^2}(u_1^2 - 2u_1 u_2 + u_2^2)x \Big|_0^L \tag{9.35}$$

---

2 Compare this to the familiar potential energy expression for a spring. Plot force, F, versus displacement, δ, using $F = k\delta$ and you will see the that the triangular area under the curve is $U = k\delta^2/2$.

Figure 9.7 The rod element with nodal forces applied.

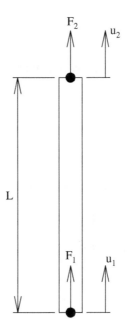

or, after evaluation at the limits,

$$U = \frac{AE}{2L}(u_1^2 - 2u_1 u_2 + u_2^2) \tag{9.36}$$

Consider now the terms that Lagrange's Equation would generate from the potential energy. For the two nodal degrees of freedom, u_1 and u_2, we will write

$$\frac{\partial U}{\partial u_1} = \frac{AE}{L}u_1 - \frac{AE}{L}u_2 \tag{9.37}$$

and

$$\frac{\partial U}{\partial u_2} = -\frac{AE}{L}u_1 + \frac{AE}{L}u_2 \tag{9.38}$$

Written together in a matrix, Equations 9.37 and 9.38 form the stiffness matrix for the element[3].

Before writing the stiffness matrix, we can apply forces, F_1 and F_2, to the nodes as shown in Figure 9.7 and find the generalized forces associated with the degrees of freedom.

If we hold u_2 constant and let u_1 vary by an amount δu_1, work will be done by F_1 but not by F_2 so we can equate

$$Q_{u_1} \delta u_1 = F_1 \delta u_1 \tag{9.39}$$

giving

$$Q_{u_1} = F_1 \tag{9.40}$$

---

3 When using Lagrange's Equation, the potential energy terms go into the stiffness matrix and the kinetic energy terms go into the mass matrix.

Similarly, varying u_2 while holding u_1 constant gives

$$Q_{u_2} = F_2 \tag{9.41}$$

We can write the stiffness terms (Equations 9.37 and 9.38) and applied force terms (Equations 9.40 and 9.41) that come from Lagrange's Equations as

$$\frac{AE}{L}u_1 - \frac{AE}{L}u_2 = F_1 \tag{9.42}$$

and

$$-\frac{AE}{L}u_1 + \frac{AE}{L}u_2 = F_2 \tag{9.43}$$

These can be put into matrix form to become

$$\begin{bmatrix} \frac{AE}{L} & -\frac{AE}{L} \\ -\frac{AE}{L} & \frac{AE}{L} \end{bmatrix} \begin{Bmatrix} u_1 \\ u_2 \end{Bmatrix} = \begin{Bmatrix} F_1 \\ F_2 \end{Bmatrix} \tag{9.44}$$

where

$$[K] = \begin{bmatrix} \frac{AE}{L} & -\frac{AE}{L} \\ -\frac{AE}{L} & \frac{AE}{L} \end{bmatrix} \tag{9.45}$$

is the element stiffness matrix.

It remains to find the mass matrix but, before doing so, we can test the finite element model of the rod statically. Consider the rod shown in Figure 9.8 where one end is connected to the ground and the other has a load, P, applied to it. If we model the rod with a single element, then we arrive at boundary conditions where $u_1 = 0$ because that end is attached to the ground and u_2 is free to move. We have the forces P at the free end and R at the constrained end. P is a known applied force and R is the reaction force provided by the ground. The equations governing the system are those from Equation 9.44 with the appropriate substitutions made into them to give

$$\begin{bmatrix} \frac{AE}{L} & -\frac{AE}{L} \\ -\frac{AE}{L} & \frac{AE}{L} \end{bmatrix} \begin{Bmatrix} 0 \\ u_2 \end{Bmatrix} = \begin{Bmatrix} R \\ P \end{Bmatrix} \tag{9.46}$$

We can extract the equations and look at them individually, yielding

$$-\frac{AE}{L}u_2 = R \tag{9.47}$$

and

$$\frac{AE}{L}u_2 = P \tag{9.48}$$

Equation 9.48 can be solved to give the displacement of the free end, u_2, as

$$u_2 = \frac{PL}{AE} \tag{9.49}$$

and this can be substituted into Equation 9.47 to find that the reaction force, R, is

$$R = -\frac{AE}{L}u_2 = -\frac{AE}{L}\frac{PL}{AE} = -P \tag{9.50}$$

Figure 9.8 A statically loaded rod modeled with one element.

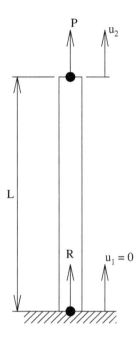

where we note that the calculated deflection exactly matches the deflection we originally found in Equation 9.18 and the reaction force exactly balances the applied force, as it should. Static finite element models always give linear algebraic equations where some of the knowns are on the left-hand side and some of the unknowns are on the right-hand side. It is simply a matter of using the equations appropriately to extract the solution.

9.3 The Mass Matrix for an Elastic Rod

Figure 9.9 shows the differential element of the finite element again but this time with mass and velocity. The mass of the differential element is $dm = \rho A dx$ where ρ is the density of the material, A is the cross-sectional area, and dx is the length of the element. The velocity of the differential element is $\dot{u}(x, t)$ where we can differentiate the shape function of Equation 9.24 to express the velocity as a function of the nodal velocities, $\dot{u}_1$ and $\dot{u}_2$, giving

$$\dot{u}(x, t) = \dot{u}_1 \left(1 - \frac{x}{L}\right) + \dot{u}_2 \left(\frac{x}{L}\right) \tag{9.51}$$

The kinetic energy of the differential element is then

$$dT = \frac{1}{2}(dm)\dot{u}^2(x, t) \tag{9.52}$$

or

$$dT = \frac{1}{2}(\rho A dx)\left[\dot{u}_1\left(1 - \frac{x}{L}\right) + \dot{u}_2\left(\frac{x}{L}\right)\right]^2 \tag{9.53}$$

The total kinetic energy of the finite element is

$$T = \int_{x=0}^{x=L} dT = \frac{1}{2}\rho A \int_0^L \left[\dot{u}_1\left(1 - \frac{x}{L}\right) + \dot{u}_2\left(\frac{x}{L}\right)\right]^2 dx \tag{9.54}$$

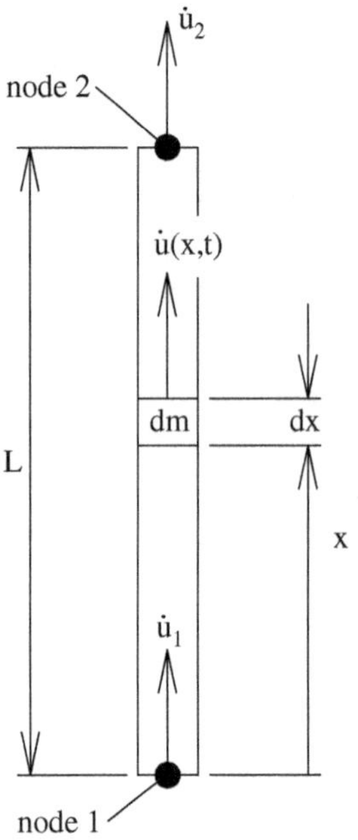

Figure 9.9   The velocity of the element of the element.

which can be expanded to

$$T = \frac{1}{2}\rho A \int_0^L \left[\dot{u}_1^2\left(1 - \frac{x}{L}\right)^2 + 2\dot{u}_1\dot{u}_2\left(1 - \frac{x}{L}\right)\left(\frac{x}{L}\right) + \dot{u}_2^2\left(\frac{x}{L}\right)^2 \right] dx \tag{9.55}$$

or

$$T = \frac{1}{2}\rho A\dot{u}_1^2 \int_0^L \left(1 - \frac{2x}{L} + \frac{x^2}{L^2}\right) dx$$

$$+ \rho A\dot{u}_1\dot{u}_2 \int_0^L \left(\frac{x}{L} - \frac{x^2}{L^2}\right) dx$$

$$+ \frac{1}{2}\rho A\dot{u}_2^2 \int_0^L \left(\frac{x^2}{L^2}\right) dx \tag{9.56}$$

Evaluating the integrals leaves us with

$$T = \frac{1}{2}\rho A\dot{u}_1^2 \left[x - \frac{2x^2}{2L} + \frac{x^3}{3L^2}\right]_0^L$$

$$+ \rho A\dot{u}_1\dot{u}_2 \left[\frac{x^2}{2L} - \frac{x^3}{3L^2}\right]_0^L$$

$$+ \frac{1}{2}\rho A\dot{u}_2^2 \left[\frac{x^3}{3L^2}\right]_0^L \tag{9.57}$$

and, after evaluating at the limits,

$$T = \frac{1}{2}\rho A\dot{u}_1^2 \left[L - L + \frac{L}{3}\right] + \rho A\dot{u}_1\dot{u}_2 \left[\frac{L}{2} - \frac{L}{3}\right] + \frac{1}{2}\rho A\dot{u}_2^2 \left[\frac{L}{3}\right] \tag{9.58}$$

which can be written as

$$T = \frac{1}{2}\rho A \dot{u}_1^2 \left(\frac{L}{3}\right) + \rho A \dot{u}_1 \dot{u}_2 \left(\frac{L}{6}\right) + \frac{1}{2}\rho A \dot{u}_2^2 \left(\frac{L}{3}\right) \tag{9.59}$$

and, finally, as

$$T = \frac{\rho AL}{6}(\dot{u}_1^2 + \dot{u}_1 \dot{u}_2 + \dot{u}_2^2) \tag{9.60}$$

The inertial terms in Lagrange's Equations of motion start with

$$\frac{\partial T}{\partial \dot{u}_1} = \frac{\rho AL}{3}\dot{u}_1 + \frac{\rho AL}{6}\dot{u}_2 \tag{9.61}$$

and

$$\frac{\partial T}{\partial \dot{u}_2} = \frac{\rho AL}{6}\dot{u}_1 + \frac{\rho AL}{3}\dot{u}_2 \tag{9.62}$$

and finish with

$$\frac{d}{dt}\left(\frac{\partial T}{\partial \dot{u}_1}\right) = \frac{\rho AL}{3}\ddot{u}_1 + \frac{\rho AL}{6}\ddot{u}_2 \tag{9.63}$$

and

$$\frac{d}{dt}\left(\frac{\partial T}{\partial \dot{u}_2}\right) = \frac{\rho AL}{6}\ddot{u}_1 + \frac{\rho AL}{3}\ddot{u}_2 \tag{9.64}$$

We now have a mass matrix to go with the stiffness matrix of Equation 9.45. The mass matrix is

$$[M] = \begin{bmatrix} \frac{\rho AL}{3} & \frac{\rho AL}{6} \\ \\ \frac{\rho AL}{6} & \frac{\rho AL}{3} \end{bmatrix} \tag{9.65}$$

and adding the inertial terms to the stiffness terms in Equation 9.44 gives the equations of motion for the rod element as

$$\begin{bmatrix} \frac{\rho AL}{3} & \frac{\rho AL}{6} \\ \\ \frac{\rho AL}{6} & \frac{\rho AL}{3} \end{bmatrix}\begin{Bmatrix} \ddot{u}_1 \\ \\ \ddot{u}_2 \end{Bmatrix} + \begin{bmatrix} \frac{AE}{L} & -\frac{AE}{L} \\ \\ -\frac{AE}{L} & \frac{AE}{L} \end{bmatrix}\begin{Bmatrix} u_1 \\ \\ u_2 \end{Bmatrix} = \begin{Bmatrix} F_1 \\ \\ F_2 \end{Bmatrix} \tag{9.66}$$

We can consider an example where the base is fixed to the ground ($u_1 = \ddot{u}_1 = 0$) and there is no load applied to the free end of the element ($F_2 = 0$) and we find

$$\begin{bmatrix} \frac{\rho AL}{3} & \frac{\rho AL}{6} \\ \\ \frac{\rho AL}{6} & \frac{\rho AL}{3} \end{bmatrix}\begin{Bmatrix} 0 \\ \\ \ddot{u}_2 \end{Bmatrix} + \begin{bmatrix} \frac{AE}{L} & -\frac{AE}{L} \\ \\ -\frac{AE}{L} & \frac{AE}{L} \end{bmatrix}\begin{Bmatrix} 0 \\ \\ u_2 \end{Bmatrix} = \begin{Bmatrix} F_1 \\ \\ 0 \end{Bmatrix} \tag{9.67}$$

The first equation in this set is

$$\frac{\rho AL}{6}\ddot{u}_2 - \frac{AE}{L}u_2 = F_1 \tag{9.68}$$

and the second is

$$\frac{\rho AL}{3}\ddot{u}_2 + \frac{AE}{L}u_2 = 0 \tag{9.69}$$

Equation 9.69 is the free vibration equation for longitudinal vibrations in the rod and can be rewritten as

$$\ddot{u}_2 + \frac{3E}{\rho L^2}u_2 = 0 \tag{9.70}$$

from which we can find the natural frequency for longitudinal rod vibrations as

$$\omega_n = \sqrt{\frac{3E}{\rho L^2}} = \frac{1}{L}\sqrt{\frac{3E}{\rho}} \tag{9.71}$$

Equation 9.68 can be used to find the reaction force, F_1, if the motion of the rod end, as expressed by $\ddot{u}_2$ and u_2, is known.

9.4 Using Multiple Elements

Figure 9.10 shows a structure constructed from two rod elements. The element numbers are shown in circles on the elements. The rod elements have two degrees of freedom each but, since one node is shared by the two elements, the system has a total of three degrees of freedom, u_1 through u_3.

Figure 9.11 shows free body diagrams of the individual elements and the nodes. The reason for separating the nodes, numbered n_1 to n_3, from the elements is to distinguish between external forces applied to the nodes, F_1 to F_3, and the internal forces acting on the elements, f_1, f_{12}, f_{21}, and f_3.

We start with a look at node n_1 where we can see that a force balance yields

$$f_1 = F_1 \tag{9.72}$$

and transferring this to the left-hand end of element 1 shows that the externally applied force, F_1, will act there just as is shown in Figure 9.10.

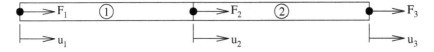

Figure 9.10 Two rod elements in an assembly.

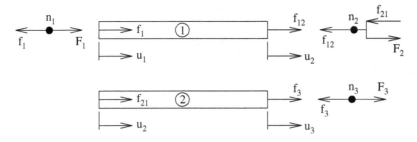

Figure 9.11 Free body diagrams of the two rod elements.

We see a similar relationship at node n_3, where

$$f_3 = F_3 \tag{9.73}$$

and again a transfer to the right end of element 2 shows the externally applied force, F_3, acting there just as is shown in Figure 9.10.

The node that is different is node n_2 where there are three forces acting. F_2 is the externally applied force at the node, f_{12} is the force that the node transfers to the right end of element 1, and f_{21} is the force that the node transfers to the left end of element 2. A force balance on the node gives

$$F_2 = f_{12} + f_{21} \tag{9.74}$$

At this point we introduce the element equations of motion for elements 1 and 2. We derived element mass and stiffness matrices in the previous section and we can use them here. For element 1 we can write Equation 9.66 as

$$\begin{bmatrix} m_{11_1} & m_{12_1} \\ m_{21_1} & m_{22_1} \end{bmatrix} \begin{Bmatrix} \ddot{u}_1 \\ \ddot{u}_2 \end{Bmatrix} + \begin{bmatrix} k_{11_1} & k_{12_1} \\ k_{21_1} & k_{22_1} \end{bmatrix} \begin{Bmatrix} u_1 \\ u_2 \end{Bmatrix} = \begin{Bmatrix} f_1 \\ f_{12} \end{Bmatrix} \tag{9.75}$$

where the notation m_{ij_1} and k_{ij_1} is used to distinguish the mass and stiffness matrix elements of finite element 1 from those of finite element 2 which will be written as m_{ij_2} and k_{ij_2} because the elements may have different properties from each other.

For element 2, we find

$$\begin{bmatrix} m_{11_2} & m_{12_2} \\ m_{21_2} & m_{22_2} \end{bmatrix} \begin{Bmatrix} \ddot{u}_2 \\ \ddot{u}_3 \end{Bmatrix} + \begin{bmatrix} k_{11_2} & k_{12_2} \\ k_{21_2} & k_{22_2} \end{bmatrix} \begin{Bmatrix} u_2 \\ u_3 \end{Bmatrix} = \begin{Bmatrix} f_{21} \\ f_3 \end{Bmatrix} \tag{9.76}$$

Consider the two matrix equations (Equations 9.75 and 9.76) as four separate equations. Equation 9.75 can be written as

$$m_{11_1} \ddot{u}_1 + m_{12_1} \ddot{u}_2 + k_{11_1} u_1 + k_{12_1} u_2 = f_1 \tag{9.77}$$

and

$$m_{21_1} \ddot{u}_1 + m_{22_1} \ddot{u}_2 + k_{21_1} u_1 + k_{22_1} u_2 = f_{12} \tag{9.78}$$

and Equation 9.76 can be written as

$$m_{11_2} \ddot{u}_2 + m_{12_2} \ddot{u}_3 + k_{11_2} u_2 + k_{12_2} u_3 = f_{21} \tag{9.79}$$

and

$$m_{21_2} \ddot{u}_2 + m_{22_2} \ddot{u}_3 + k_{21_2} u_2 + k_{22_2} u_3 = f_3 \tag{9.80}$$

We now make use of Equations 9.72, 9.73, and 9.74 to change the internal forces to external forces. Equations 9.72 and 9.73 let us immediately replace the right-hand sides of Equations 9.77 and 9.80 with the externally applied forces, F_1 and F_3. Equation 9.74 on the other hand indicates that we must add together f_{12} and f_{21} before they can be replaced by the externally applied force, F_2. We do this by adding together Equations 9.78 and 9.79 to get

$$m_{21_1} \ddot{u}_1 + (m_{22_1} + m_{11_2}) \ddot{u}_2 + m_{12_2} \ddot{u}_3$$
$$+ k_{21_1} u_1 + (k_{22_1} + k_{11_2}) u_2 + k_{12_2} u_3 = f_{12} + f_{21} = F_2 \tag{9.81}$$

Combining Equations 9.78 and 9.79 leaves us with just three equations for the three nodal degrees of freedom shown in Figure 9.10 as we should have expected. Written in matrix form, they are

$$
\begin{bmatrix} m_{11_1} & m_{12_1} & 0 \\ m_{21_1} & (m_{22_1} + m_{11_2}) & m_{12_2} \\ 0 & m_{21_2} & m_{22_2} \end{bmatrix} \begin{Bmatrix} \ddot{u}_1 \\ \ddot{u}_2 \\ \ddot{u}_3 \end{Bmatrix}
$$

$$
+ \begin{bmatrix} k_{11_1} & k_{12_1} & 0 \\ k_{21_1} & (k_{22_1} + k_{11_2}) & k_{12_2} \\ 0 & k_{21_2} & k_{22_2} \end{bmatrix} \begin{Bmatrix} u_1 \\ u_2 \\ u_3 \end{Bmatrix} = \begin{Bmatrix} F_1 \\ F_2 \\ F_3 \end{Bmatrix} \tag{9.82}
$$

where it is clear that we could generate the 3×3 *global mass matrix* by overlapping the two 2×2 finite element mass matrices inside the 3×3 matrix and adding the terms together where they share degrees of freedom. Matrix elements outside the overlapping finite element matrices are set to zero since they bring in nodal degrees of freedom that are not coupled to the degree of freedom whose equation of motion is represented by the row being considered. The *global stiffness matrix* is assembled in the same way.

Equation 9.82 can be expanded to the general case where we assemble a complete finite element model from many elements. The *global mass matrix* is shown in Figure 9.12 for the case where there are N elements. As with the two rod elements, the global mass matrix is formed by writing the N finite element mass matrices and then superimposing them inside the global matrix where the elements of the mass matrices that share nodal degrees of freedom are added together (the $+$ signs) to combine equations of motion. Large parts of the matrix where degrees of freedom are not coupled to the degree of freedom whose equation is being written are filled with zeroes.

The *global stiffness matrix* is found in the same way and is shown in Figure 9.13.

This procedure applies to every type of finite element, not just to rod elements. Mass and stiffness matrices are written for every finite element in the system and the global matrices are assembled as described here. Finally, the external nodal forces are applied and the

Figure 9.12 The global mass matrix.

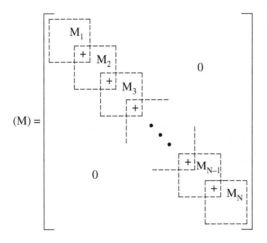

Figure 9.13 The global stiffness matrix.

$$(K) = $$

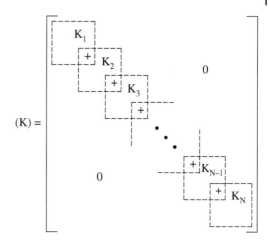

resulting set of N equations of motion for the entire structure becomes

$$[M]\{\ddot{u}\} + [K]\{u\} = \{F\} \tag{9.83}$$

We then apply boundary conditions and solve the equations as we would for any other NDOF system.

9.5 The Two-noded Beam Element

We return now to the material in Section 9.1 where we started to work on the shape function for a uniform beam before going off to examine the simpler case of the elastic rod. We determined there (see Equation 9.13) that the deflected shape of a uniform beam would be a cubic function of position along the beam. We left some undetermined coefficients but we will determine them now.

Consider the beam element shown in Figure 9.14. The element has four degrees of freedom – the deflection at node 1, u_1, the slope at node 1, u_2, the deflection at node 2, u_3, and the slope at node 2, u_4. The positive direction for deflections, $y(x)$, is upwards and the position within the element, x, is measured positively to the right. As a result, the slopes, $\theta = dy/dx$, will be positive in a counterclockwise direction. All of the degrees of freedom are shown in their positive directions on the figure.

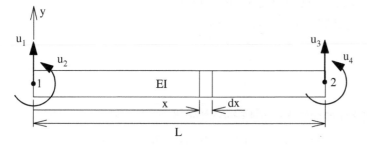

Figure 9.14 The two-noded beam element.

Let the general cubic shape of the deflected element be

$$y(x) = a_0 + a_1 x + a_2 x^2 + a_3 x^3 \qquad (9.84)$$

and consider the boundary conditions. They are

$$y(0) = u_1$$

$$\frac{dy(0)}{dx} = u_2$$

$$y(L) = u_3$$

$$\frac{dy(L)}{dx} = u_4$$

Differentiating Equation 9.84 gives the general expression for the slope as

$$\frac{dy(x)}{dx} = a_1 + 2a_2 x + 3a_3 x^2 \qquad (9.85)$$

Applying the boundary conditions to Equations 9.84 and 9.85 gives the following four equations.

$$y(0) = a_0 = u_1 \qquad (9.86)$$

$$\frac{dy(0)}{dx} = a_1 = u_2 \qquad (9.87)$$

$$y(L) = a_0 + a_1 L + a_2 L^2 + a_3 L^3 = u_3 \qquad (9.88)$$

$$\frac{dy(L)}{dx} = a_1 + 2a_2 L + 3a_3 L^2 = u_4 \qquad (9.89)$$

Equations 9.86 to 9.89 are sufficient to solve for the four unknown coefficients, $a_0, a_1, a_2,$ and a_3, in terms of the nodal degrees of freedom. Solving yields

$$a_0 = u_1$$

$$a_1 = u_2$$

$$a_2 = -\frac{3}{L^2} u_1 - \frac{2}{L} u_2 + \frac{3}{L^2} u_3 - \frac{1}{L} u_4$$

$$a_3 = \frac{2}{L^3} u_1 + \frac{1}{L^2} u_2 - \frac{2}{L^3} u_3 + \frac{1}{L^2} u_4$$

We can use the values of the coefficients to express $y(x)$ as

$$y(x) = \left(1 - \frac{3x^2}{L^2} + \frac{2x^3}{L^3}\right) u_1 + \left(x - \frac{2x^2}{L} + \frac{x^3}{L^2}\right) u_2$$

$$+ \left(\frac{3x^2}{L^2} - \frac{2x^3}{L^3}\right) u_3 + \left(-\frac{x^2}{L} + \frac{x^3}{L^2}\right) u_4 \qquad (9.90)$$

9.5.1 The Two-noded Beam Element – Stiffness Matrix

In the same way that we considered the potential energy of the elastic rod in Equations 9.28 to 9.30, we can look at the energy in the differential element of the beam shown in

Figure 9.15 *M versus dθ for the element.*

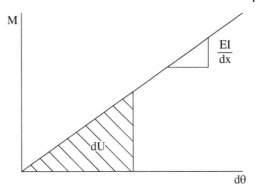

Figure 9.14. The constitutive relationship for the beam bending element was first given in Equation 9.7 where we said that the plus or minus sign in the expression depended on the positive directions chosen for x and y. For the directions chosen here, the plus sign is appropriate, so we can write

$$M = EI\frac{d^2y}{dx^2} \tag{9.91}$$

Note that the second derivative of y with respect to x^2 is equal to the first derivative of θ with respect to x, where θ is the slope dy/dx. We can therefore write the bending constitutive relationship as

$$M = EI\frac{d\theta}{dx} \tag{9.92}$$

or as

$$M = \left(\frac{EI}{dx}\right)d\theta \tag{9.93}$$

which is plotted in Figure 9.15. The product of a moment and an angle results in work or energy so the area under the curve in the figure can be interpreted as the work done when a moment causes a beam deflection or as the potential energy stored in the beam as it is forced to deflect. As it was with the rod element or a linear spring, the area under the curve is triangular and we can express it as one half its base, $d\theta$, multiplied by its height, $M = (EI/dx)d\theta$, to get

$$dU = \frac{1}{2}(d\theta)\left(\frac{EI}{dx}\right)d\theta \tag{9.94}$$

which can be multiplied and divided by dx and rearranged to get

$$dU = \frac{1}{2}EI\left(\frac{d\theta}{dx}\right)^2 dx \tag{9.95}$$

or finally, replacing θ with dy/dx,

$$dU = \frac{1}{2}EI\left(\frac{d^2y}{dx^2}\right)^2 dx \tag{9.96}$$

The total potential energy is then found by integrating over the length of the beam element and is

$$U = \frac{1}{2}EI\int_0^L \left(\frac{d^2y}{dx^2}\right)^2 dx \tag{9.97}$$

Using Lagrange's Equation, the terms that make up the stiffness matrix will come from

$$\frac{\partial U}{\partial u_i} = EI \int_0^L \left(\frac{d^2y}{dx^2}\right) \frac{\partial}{\partial u_i}\left(\frac{d^2y}{dx^2}\right) dx; i = 1,4 \tag{9.98}$$

We can differentiate Equation 9.90 twice to get

$$\frac{d^2y}{dx^2} = \left(-\frac{6}{L^2} + \frac{12x}{L^3}\right) u_1 + \left(-\frac{4}{L} + \frac{6x}{L^2}\right) u_2$$

$$+ \left(\frac{6}{L^2} - \frac{12x}{L^3}\right) u_3 + \left(-\frac{2}{L} + \frac{6x}{L^2}\right) u_4 \tag{9.99}$$

Then, for u_1, we find

$$\frac{\partial}{\partial u_1}\left(\frac{d^2y}{dx^2}\right) = \left(-\frac{6}{L^2} + \frac{12x}{L^3}\right) \tag{9.100}$$

Equation 9.98 can then be written as

$$\frac{\partial U}{\partial u_1} = EI \left[\int_0^L \left(-\frac{6}{L^2} + \frac{12x}{L^3}\right)\left(-\frac{6}{L^2} + \frac{12x}{L^3}\right) dx\right] u_1$$

$$+ EI \left[\int_0^L \left(-\frac{4}{L} + \frac{6x}{L^2}\right)\left(-\frac{6}{L^2} + \frac{12x}{L^3}\right) dx\right] u_2$$

$$+ EI \left[\int_0^L \left(\frac{6}{L^2} - \frac{12x}{L^3}\right)\left(-\frac{6}{L^2} + \frac{12x}{L^3}\right) dx\right] u_3$$

$$+ EI \left[\int_0^L \left(-\frac{2}{L} + \frac{6x}{L^2}\right)\left(-\frac{6}{L^2} + \frac{12x}{L^3}\right) dx\right] u_4 \tag{9.101}$$

The polynomials inside the square brackets can be multiplied to give

$$\frac{\partial U}{\partial u_1} = EI \left[\int_0^L \left(\frac{36}{L^4} - \frac{144x}{L^5} + \frac{144x^2}{L^6}\right) dx\right] u_1$$

$$+ EI \left[\int_0^L \left(\frac{24}{L^3} - \frac{84x}{L^4} + \frac{72x^2}{L^5}\right) dx\right] u_2$$

$$+ EI \left[\int_0^L \left(-\frac{36}{L^4} + \frac{144x}{L^5} - \frac{144x^2}{L^6}\right) dx\right] u_3$$

$$+ EI \left[\int_0^L \left(\frac{12}{L^3} - \frac{60x}{L^4} + \frac{72x^2}{L^5}\right) dx\right] u_4 \tag{9.102}$$

The integrals of the polynomials are relatively easy to evaluate, yielding

$$\frac{\partial U}{\partial u_1} = EI \left[\frac{36x}{L^4} - \frac{144x^2}{2L^5} + \frac{144x^3}{3L^6}\right]_0^L u_1$$

$$+ EI \left[\frac{24x}{L^3} - \frac{84x^2}{2L^4} + \frac{72x^3}{3L^5}\right]_0^L u_2$$

$$+ EI \left[-\frac{36x}{L^4} + \frac{144x^2}{2L^5} - \frac{144x^3}{3L^6}\right]_0^L u_3$$

$$+ EI \left[\frac{12x}{L^3} - \frac{60x^2}{2L^4} + \frac{72x^3}{3L^5}\right]_0^L u_4 \tag{9.103}$$

and, after evaluating them at the limits, we have

$$\frac{\partial U}{\partial u_1} = \left(\frac{12EI}{L^3}\right)u_1 + \left(\frac{6EI}{L^2}\right)u_2 + \left(\frac{-12EI}{L^3}\right)u_3 + \left(\frac{6EI}{L^2}\right)u_4 \qquad (9.104)$$

which is the first row out of the 4×4 stiffness matrix for the two-noded beam element.

Following the same procedure for the three remaining degrees of freedom leads to the stiffness matrix

$$[K] = \frac{EI}{L^3}\begin{bmatrix} 12 & 6L & -12 & 6L \\ 6L & 4L^2 & -6L & 2L^2 \\ -12 & -6L & 12 & -6L \\ 6L & 2L^2 & -6L & 4L^2 \end{bmatrix} \qquad (9.105)$$

9.5.2 The Two-noded Beam Element – Mass Matrix

In order to derive the mass matrix using Lagrange's Equations, we need to have an expression for the total kinetic energy of the finite element. Consider the infinitesimal element shown in Figure 9.14. If the beam has constant mass per unit length, m, within the finite element, then the mass of the infinitesimal element of length dx is $dm = m\,dx$. We have an expression for the deflection of the finite element at the point, x, where the infinitesimal element is located. It is $y(x)$ as given in Equation 9.90. Since we are now considering inertial effects, we need to expand the definition so that it becomes a function of time as well. We therefore write

$$y(x,t) = \left(1 - \frac{3x^2}{L^2} + \frac{2x^3}{L^3}\right)u_1(t) + \left(x - \frac{2x^2}{L} + \frac{x^3}{L^2}\right)u_2(t)$$
$$+ \left(\frac{3x^2}{L^2} - \frac{2x^3}{L^3}\right)u_3(t) + \left(-\frac{x^2}{L} + \frac{x^3}{L^2}\right)u_4(t) \qquad (9.106)$$

where it is recognized that the element degrees of freedom are time-varying. The velocity of the infinitesimal element is then

$$\frac{\partial y(x,t)}{\partial t} = \left(1 - \frac{3x^2}{L^2} + \frac{2x^3}{L^3}\right)\dot{u}_1(t) + \left(x - \frac{2x^2}{L} + \frac{x^3}{L^2}\right)\dot{u}_2(t)$$
$$+ \left(\frac{3x^2}{L^2} - \frac{2x^3}{L^3}\right)\dot{u}_3(t) + \left(-\frac{x^2}{L} + \frac{x^3}{L^2}\right)\dot{u}_4(t) \qquad (9.107)$$

and the contribution of the infinitesimal element to the kinetic energy is

$$dT = \frac{1}{2}(m\,dx)\left(\frac{\partial y(x,t)}{\partial t}\right)^2 = \frac{1}{2}m\left(\frac{\partial y(x,t)}{\partial t}\right)^2 dx \qquad (9.108)$$

The total kinetic energy of the finite element is then

$$T = \frac{1}{2}m\int_0^L \left(\frac{\partial y(x,t)}{\partial t}\right)^2 dx \qquad (9.109)$$

The inertial terms that enter Lagrange's equations of motion are

$$\frac{d}{dt}\left(\frac{\partial T}{\partial \dot{u}_i}\right) - \left(\frac{\partial T}{\partial u_i}\right); i = 1, 4 \qquad (9.110)$$

Consideration of Equation 9.107 shows that the velocity is not a function of the nodal displacements so that all terms of the form $(\partial T/\partial u_i)$ will be zero, leaving the inertial terms as

$$\frac{d}{dt}\left(\frac{\partial T}{\partial \dot{u}_i}\right); i = 1, 4 \tag{9.111}$$

Differentiating Equation 9.109 with respect to $\dot{u}_i$ yields

$$\frac{\partial T}{\partial \dot{u}_i} = m \int_0^L \left(\frac{\partial y(x,t)}{\partial t}\right) \frac{\partial}{\partial \dot{u}_i}\left(\frac{\partial y(x,t)}{\partial t}\right) dx \tag{9.112}$$

We now follow the same procedure we did for the stiffness matrix.
For example, for $\dot{u}_1$, we differentiate equation 9.107 to find

$$\frac{\partial}{\partial \dot{u}_1}\left(\frac{\partial y(x,t)}{\partial t}\right) = \left(1 - \frac{3x^2}{L^2} + \frac{2x^3}{L^3}\right) \tag{9.113}$$

so that Equations 9.107 and 9.113 can be substituted into Equation 9.112 and it can be written as

$$\frac{\partial T}{\partial \dot{u}_i} = m \left[\int_0^L \left(1 - \frac{3x^2}{L^2} + \frac{2x^3}{L^3}\right)\left(1 - \frac{3x^2}{L^2} + \frac{2x^3}{L^3}\right) dx\right] \dot{u}_1$$

$$+ m \left[\int_0^L \left(x - \frac{2x^2}{L} + \frac{x^3}{L^2}\right)\left(1 - \frac{3x^2}{L^2} + \frac{2x^3}{L^3}\right) dx\right] \dot{u}_2$$

$$+ m \left[\int_0^L \left(\frac{3x^2}{L^2} - \frac{2x^3}{L^3}\right)\left(1 - \frac{3x^2}{L^2} + \frac{2x^3}{L^3}\right) dx\right] \dot{u}_3$$

$$+ m \left[\int_0^L \left(-\frac{x^2}{L} + \frac{x^3}{L^2}\right)\left(1 - \frac{3x^2}{L^2} + \frac{2x^3}{L^3}\right) dx\right] \dot{u}_4 \tag{9.114}$$

Multiplying the polynomials yields

$$\frac{\partial T}{\partial \dot{u}_i} = m \left[\int_0^L \left(1 - \frac{6x^2}{L^2} + \frac{4x^3}{L^3} + \frac{9x^4}{L^4} - \frac{12x^5}{L^5} + \frac{4x^6}{L^6}\right) dx\right] \dot{u}_1$$

$$+ m \left[\int_0^L \left(x - \frac{2x^2}{L} - \frac{2x^3}{L^2} + \frac{8x^4}{L^3} - \frac{7x^5}{L^4} + \frac{2x^6}{L^5}\right) dx\right] \dot{u}_2$$

$$+ m \left[\int_0^L \left(\frac{3x^2}{L^2} - \frac{2x^3}{L^3} - \frac{9x^4}{L^4} + \frac{12x^5}{L^5} - \frac{4x^6}{L^6}\right) dx\right] \dot{u}_3$$

$$+ m \left[\int_0^L \left(-\frac{x^2}{L} + \frac{x^3}{L^2} + \frac{3x^4}{L^3} - \frac{5x^5}{L^4} + \frac{2x^6}{L^5}\right) dx\right] \dot{u}_4 \tag{9.115}$$

Integrating gives

$$\frac{\partial T}{\partial \dot{u}_i} = m \left[x - \frac{6x^3}{3L^2} + \frac{4x^4}{4L^3} + \frac{9x^5}{5L^4} - \frac{12x^6}{6L^5} + \frac{4x^7}{7L^6}\right]_0^L \dot{u}_1$$

$$+ m \left[\frac{x^2}{2} - \frac{2x^3}{3L} - \frac{2x^4}{4L^2} + \frac{8x^5}{5L^3} - \frac{7x^6}{6L^4} + \frac{2x^7}{7L^5}\right]_0^L \dot{u}_2$$

$$+ m \left[\frac{3x^3}{3L^2} - \frac{2x^4}{4L^3} - \frac{9x^5}{5L^4} + \frac{12x^6}{6L^5} - \frac{4x^7}{7L^6} \right]_0^L \dot{u}_3$$

$$+ m \left[-\frac{x^3}{3L} + \frac{x^4}{4L^2} + \frac{3x^5}{5L^3} - \frac{5x^6}{6L^4} + \frac{2x^7}{7L^5} \right]_0^L \dot{u}_4 \tag{9.116}$$

and, evaluating at the limits, we find

$$\frac{\partial T}{\partial \dot{u}_i} = \left(\frac{156mL}{420} \right) \dot{u}_1 + \left(\frac{22mL^2}{420} \right) \dot{u}_2 + \left(\frac{54mL}{420} \right) \dot{u}_3 + \left(\frac{-13mL^2}{420} \right) \dot{u}_4 \tag{9.117}$$

Equation 9.117 contains the elements in the first row of the mass matrix for the element. If we repeat the procedure for the remaining three degrees of freedom, we find that the mass matrix is

$$[M] = \frac{mL}{420} \begin{bmatrix} 156 & 22L & 54 & -13L \\ 22L & 4L^2 & 13L & -3L^2 \\ 54 & 13L & 156 & -22L \\ -13L & -3L^2 & -22L & 4L^2 \end{bmatrix} \tag{9.118}$$

As with the rod elements in the previous section, the global mass and stiffness matrices for N finite elements are formed by superimposing the element matrices along the main diagonal of the global matrices, adding the elements together where they share degrees of freedom, and filling the remainder of the matrices with zeros. See Figures 9.12 and 9.13.

9.6 Two-noded Beam Element Vibrations Example

The following is a step-by-step example of formulating a finite element model for beam vibrations. While the example has only two elements, the procedure is the same for any number of elements. The example uses 2-noded beam elements with 2 degrees-of-freedom per node to model a uniform beam of length L having a mass per unit length of m and a flexural rigidity EI that remains constant over the entire beam.

1. Choose node and element numbers. Pay no attention to boundary conditions at this point. The nodes are numbered 1 to 3 and the element numbers are in the circles in Figure 9.16.
2. Write the mass and stiffness matrices for each element. The degrees-of-freedom are the translations of a node perpendicular to the beam and the rotation at the node. The mass matrix for an element is

$$[M]_i = \frac{m\ell}{420} \begin{bmatrix} 156 & 22\ell & 54 & -13\ell \\ 22\ell & 4\ell^2 & 13\ell & -3\ell^2 \\ 54 & 13\ell & 156 & -22\ell \\ -13\ell & -3\ell^2 & -22\ell & 4\ell^2 \end{bmatrix}$$

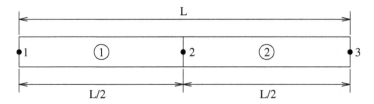

Figure 9.16 Node and element numbering.

For these elements, $\ell = L/2$ so the elemental mass matrix for each element is:

$$[M]_1 = [M]_2 = \frac{mL}{3360} \begin{bmatrix} 624 & 44L & 216 & -26L \\ 44L & 4L^2 & 26L & -3L^2 \\ 216 & 26L & 624 & -44L \\ -26L & -3L^2 & -44L & 4L^2 \end{bmatrix}$$

Similarly, the stiffness matrix for a 4DOF, 2-noded beam element is:

$$[K]_i = \frac{EI}{\ell^3} \begin{bmatrix} 12 & 6\ell & -12 & 6\ell \\ 6\ell & 4\ell^2 & -6\ell & 2\ell^2 \\ -12 & -6\ell & 12 & -6\ell \\ 6\ell & 2\ell^2 & -6\ell & 4\ell^2 \end{bmatrix}$$

And, again substituting $\ell = L/2$ for each element, the elemental stiffness matrices are

$$[K]_1 = [K]_2 = \frac{4EI}{L^3} \begin{bmatrix} 24 & 6L & -24 & 6L \\ 6L & 2L^2 & -6L & L^2 \\ -24 & -6L & 24 & -6L \\ 6L & L^2 & -6L & 2L^2 \end{bmatrix}$$

3. Define the degrees-of-freedom for the entire structure by naming the translations and rotations at the nodes. For this example, we define a vector of degrees-of-freedom $\{u\}$ as follows

$$\begin{Bmatrix} u_1 \\ u_2 \\ u_3 \\ u_4 \\ u_5 \\ u_6 \end{Bmatrix} = \begin{Bmatrix} \text{translation of node 1} \\ \text{rotation of node 1} \\ \text{translation of node 2} \\ \text{rotation of node 2} \\ \text{translation of node 3} \\ \text{rotation of node 3} \end{Bmatrix}$$

Figure 9.17 shows the degrees-of-freedom.
4. Define the externally applied loads, both forces and moments, at the nodes. The forces and moments correspond exactly with the degrees-of-freedom as to numbering system

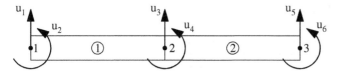

Figure 9.17   Node and element numbering.

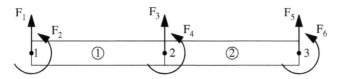

Figure 9.18   Externally applied forces and moments.

and positive directions. The forces and moments are kept in the vector $\{F\}$, defined as follows

$$
\begin{Bmatrix} F_1 \\ F_2 \\ F_3 \\ F_4 \\ F_5 \\ F_6 \end{Bmatrix} = \begin{Bmatrix} \text{force applied at node 1} \\ \text{moment applied at node 1} \\ \text{force applied at node 2} \\ \text{moment applied at node 2} \\ \text{force applied at node 3} \\ \text{moment applied at node 3} \end{Bmatrix}
$$

Figure 9.18 shows the externally applied forces and moments.

5. Assemble the global mass and stiffness matrices.

$$
[M] = \frac{mL}{3360} \begin{bmatrix}
624 & 44L & 216 & -26L & 0 & 0 \\
44L & 4L^2 & 26L & -3L^2 & 0 & 0 \\
216 & 26L & 624+624=1248 & -44L+44L=0 & 216 & -26L \\
-26L & -3L^2 & -44L+44L=0 & 4L^2+4L^2=8L^2 & 26L & -3L^2 \\
0 & 0 & 216 & 26L & 624 & -44L \\
0 & 0 & -26L & -3L^2 & -44L & 4L^2
\end{bmatrix}
$$

$$
[K] = \frac{4EI}{L^3} \begin{bmatrix}
24 & 6L & -24 & 6L & 0 & 0 \\
6L & 2L^2 & -6L & L^2 & 0 & 0 \\
-24 & -6L & 24+24=48 & -6L+6L=0 & -24 & 6L \\
6L & L^2 & -6L+6L=0 & 2L^2+2L^2=4L^2 & -6L & L^2 \\
0 & 0 & -24 & -6L & 24 & -6L \\
0 & 0 & 6L & L^2 & -6L & 2L^2
\end{bmatrix}
$$

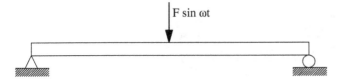

Figure 9.19 The harmonically varying applied load.

6. Write the global equations of motion.

$$\frac{mL}{3360}\begin{bmatrix} 624 & 44L & 216 & -26L & 0 & 0 \\ 44L & 4L^2 & 26L & -3L^2 & 0 & 0 \\ 216 & 26L & 1248 & 0 & 216 & -26L \\ -26L & -3L^2 & -0 & 8L^2 & 26L & -3L^2 \\ 0 & 0 & 216 & 26L & 624 & -44L \\ 0 & 0 & -26L & -3L^2 & -44L & 4L^2 \end{bmatrix}\begin{Bmatrix} \ddot{u}_1 \\ \ddot{u}_2 \\ \ddot{u}_3 \\ \ddot{u}_4 \\ \ddot{u}_5 \\ \ddot{u}_6 \end{Bmatrix}$$

$$+\frac{4EI}{L^3}\begin{bmatrix} 24 & 6L & -24 & 6L & 0 & 0 \\ 6L & 2L^2 & -6L & L^2 & 0 & 0 \\ -24 & -6L & 48 & 0 & -24 & 6L \\ 6L & L^2 & 0 & 4L^2 & -6L & L^2 \\ 0 & 0 & -24 & -6L & 24 & -6L \\ 0 & 0 & 6L & L^2 & -6L & 2L^2 \end{bmatrix}\begin{Bmatrix} u_1 \\ u_2 \\ u_3 \\ u_4 \\ u_5 \\ u_6 \end{Bmatrix}=\begin{Bmatrix} F_1 \\ F_2 \\ F_3 \\ F_4 \\ F_5 \\ F_6 \end{Bmatrix}$$

Note the standard form for a linear vibrations problem. That is, $[M]\{\ddot{u}\}+[K]\{u\}=\{F\}$.

7. Apply the boundary conditions and forcing functions. For this example, let the beam be pinned at each end and be subjected to a harmonically varying load at its center as shown in Figure 9.19.

That is $u_1 = u_5 = 0$ – zero displacements at the ends (nodes 1 and 3). Note that u_2 and u_6 are not zero because the pinned joints at the ends allow rotations. $\ddot{u}_1 = \ddot{u}_5 = 0$ – since there are no displacements, there cannot be any accelerations for these degrees-of-freedom. $F_2 = F_4 = F_6 = 0$ since there are no applied moments. $F_3 = -F \sin \omega t$ – this is the applied load with a known magnitude F and frequency ω and it is acting in the negative sense of the degree-of-freedom. F_1 and F_5 are unknown reaction forces.

The first and fifth equations can be extracted from the global set since they are only useful for finding the reaction forces. Note that columns 1 and 5 have been removed from the global equations because $u_1, u_5, \ddot{u}_1$, and $\ddot{u}_5$ are all zero.

$$\frac{mL}{3360}\begin{bmatrix} 44L & 216 & -26L & 0 \\ 0 & 216 & 26L & -44L \end{bmatrix}\begin{Bmatrix} \ddot{u}_2 \\ \ddot{u}_3 \\ \ddot{u}_4 \\ \ddot{u}_6 \end{Bmatrix}$$

$$+\frac{4EI}{L^3}\begin{bmatrix} 6L & -24 & 6L & 0 \\ 0 & -24 & -6L & -6L \end{bmatrix}\begin{Bmatrix} u_2 \\ u_3 \\ u_4 \\ u_6 \end{Bmatrix} = \begin{Bmatrix} F_1 \\ F_5 \end{Bmatrix}$$

Once the displacements and accelerations have been found, these equations can be used to find the reaction forces, F_1 and F_5.

8. The remaining equations are the equations-of-motion for the system. Notice that the mass and stiffness matrices are found by deleting the rows and columns associated with the boundary conditions (i.e. rows 1 and 5 and columns 1 and 5).

$$\frac{mL}{3360}\begin{bmatrix} 4L^2 & 26L & -3L^2 & 0 \\ 26L & 1248 & 0 & -26L \\ -3L^2 & -0 & 8L^2 & -3L^2 \\ 0 & -26L & -3L^2 & 4L^2 \end{bmatrix}\begin{Bmatrix} \ddot{u}_2 \\ \ddot{u}_3 \\ \ddot{u}_4 \\ \ddot{u}_6 \end{Bmatrix}$$

$$+\frac{4EI}{L^3}\begin{bmatrix} 2L^2 & -6L & L^2 & 0 \\ -6L & 48 & 0 & 6L \\ L^2 & 0 & 4L^2 & L^2 \\ 0 & 6L & L^2 & 2L^2 \end{bmatrix}\begin{Bmatrix} u_2 \\ u_3 \\ u_4 \\ u_6 \end{Bmatrix} = \begin{Bmatrix} 0 \\ -F\sin\omega t \\ 0 \\ 0 \end{Bmatrix}$$

9. Assume harmonic motion at the forcing frequency ω and solve as usual. That is, assume

$$\begin{Bmatrix} u_2 \\ u_3 \\ u_4 \\ u_6 \end{Bmatrix} = \begin{Bmatrix} A_1 \\ A_2 \\ A_3 \\ A_4 \end{Bmatrix}\sin\omega t$$

and solve for the amplitudes of the degrees of freedom. Alternatively, the case where $F = 0$ could be considered and the eigenvalues and eigenvectors would provide the natural frequencies and mode shapes for the system.

Exercises

9.1 Analytical expressions for the natural frequencies of continuous, uniform, cantilever beams were developed at the end of Chapter 8 (see Equation 8.87). Use the finite element technique with a single two-noded beam element to get an approximation for the first two natural frequencies of a uniform cantilever beam. How do these compare to those found in Chapter 8?

9.2 Writing and solving the continuous differential equation governing longitudinal vibrations in a uniform rod yields natural frequencies as a function of the properties and dimensions of the rod and its boundary conditions. For a rod that is fixed at one end and free at the other, the natural frequencies are

$$\omega_n = \left(\frac{2n+1}{2}\right)\frac{\pi}{\ell}\sqrt{\frac{E}{\rho}} \quad n = 0, 1, 2, ...$$

where ℓ is the length of the rod, E is Young's modulus, and ρ is the density of the material.

Use three elements to construct a finite element model of the uniform rod just described and find an expression for its lowest natural frequency. What is the percentage difference between your result and that from the continuous model?

9.3 The figure shows a structure containing three uniform segments.
· Segment $A - B$ has mass, $2m$, length, L, and cross-sectional area, $2A$.
· Segment $B - C$ has mass, m, length, L, and cross-sectional area, A.
· Segment $C - D$ has mass, $2m$, length, L, and cross-sectional area, $2A$.
The entire structure is made from the same material so that Young's modulus, E, is the same everywhere.
The ends of the structure (points A and D) are fixed rigidly to the ground.
The task is to create a finite element model of the structure using rod elements in order to determine the two natural frequencies and corresponding mode shapes for longitudinal vibrations of the structure. Do the following.
(a) Use the three segments as three finite elements and write the mass and stiffness matrices for each of the three elements. (Hint: The mass matrix in this chapter is in terms of ρAL but that is the same as m as defined here. The stiffness matrix in the chapter uses AE/L but you can define $k = AE/L$ and work with it instead for efficiency.)
(b) Write the global equations of motion.
(c) Apply the boundary conditions.
(d) Find expressions for the two natural frequencies in terms of m and k as defined above.
(e) Find the two mode shapes.

9.4 A concrete column supporting a railway bridge is constructed with a step change in cross-sectional area at its center as shown in the figure. As trains travel over the bridge, the vertical forces from the wheels are transmitted to the column and produce a harmonically varying force, $F \sin \omega t$, with a frequency that can be calculated from the distance between axles and the speed of the train. There is a concern that, because the forcing frequency may be near a natural frequency of the column, the vertical reaction force at the foundation is much greater than the wheel forces applied to the top of the column. It has been decided to create a finite element model of the column using two rod elements. The model will then be used to predict the reaction force at the foundation due to a known vertical wheel force. The smaller part of the column has total mass, M, and cross-sectional area, A. The larger part has total mass, $2M$, and cross-sectional area, $2A$. Concrete is used throughout so E is the same everywhere.
(a) Using the parameters for the column, write the mass and stiffness matrices for each of the two elements.
(b) Write the global equations of motion for the column.

Figure E9.3

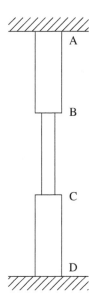

(c) Apply the boundary conditions and write the final form of the equations of motion.

(d) Explain how you would determine the magnitude of the reaction force at the foundation from your equations.

Figure E9.4

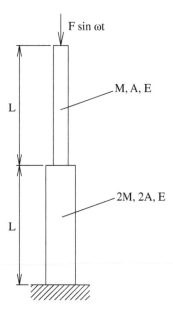

9.5 Read Subsection 9.5.1 and Section 9.6 and do the following.

(a) Show that Equation 9.98 follows from Lagrange's Equation and Equation 9.97.

(b) For the example in Section 9.6, write the equations of motion for the case where the beam is cantilevered at the left end and free at the right.

10

The Inerter

Since Newton formulated his three laws of motion, the effect of mass on a system has been related to absolute acceleration (i.e. acceleration with respect to ground). Until recently[1] , there has never been a device that delivers a force proportional to relative acceleration. Now there is such a device and it is called the *Inerter*. The inerter first saw practical use in the suspension of a Formula 1 car in 2005 and has a great deal of potential for use in systems designed for controlling vibrations.

10.1 Modeling the Inerter

The basic idea of the inerter is that the relative translational accelerations of two points on the device are converted to angular acceleration of a disk. One way to do that is shown in Figure 10.1 where we start with the device anchored to a wall to simplify the explanation. The basic parts are: a hollow tube, closed at one end and with a hole drilled through the other end; a rotating disk supported inside the tube by bearings and with a threaded hole through its center; a connector fixed to the closed end of the tube; a long threaded rod that is threaded through the disk passes through the opening at the end of the tube and has a moving connector at its free end.

Figure 10.2 shows a simple screw which we will use to explain the operation of the inerter. Figure 10.1 shows a force F being applied to the moving end of the inerter and a corresponding displacement x. Leaving aside the force for the moment, we can relate the displacement x to the rotation of the disk by considering the action of the screw. The screw has a pitch ℓ which is the distance that it moves during one complete revolution. Figure 10.2 also shows the wedge that results from unwrapping one revolution of the screw. The vertical distance traveled is ℓ. If we consider the rotation of the disk to be measured by an angle θ, then one complete revolution occurs when $\theta = 2\pi$ radians. One complete revolution also has $x = \ell$. Taking ratios of a complete rotation, we can write

$$\frac{\theta}{2\pi} = \frac{x}{\ell} \tag{10.1}$$

1 Smith, Malcolm C., *Synthesis of Mechanical Networks: The Inerter*, IEEE Transactions on Automatic Control, Vol. 47, No. 10, October 2002, pp. 1648–1662.

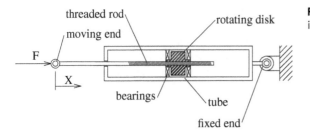

Figure 10.1 An inerter implementation.

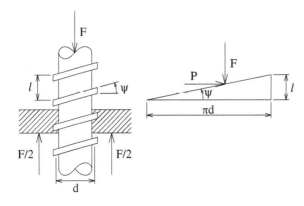

Figure 10.2 A screw.

from which

$$\theta = \frac{2\pi}{\ell}x \qquad (10.2)$$

and differentiating twice leads to

$$\ddot{\theta} = \frac{2\pi}{\ell}\ddot{x} \qquad (10.3)$$

We now consider the forces acting on the screw. The force F will be transmitted through the rod and will act on the disk as a distributed force over the thread contact area. This can be modeled as a single force acting on the unwrapped thread as shown in Figure 10.2. In order to turn the screw, the force P must be applied. P acts at a distance $d/2$ from the center line of the screw where d is the mean thread diameter. In the absence of friction, we can sum forces along the plane of the screw and write

$$P\cos\psi - F\sin\psi = 0 \qquad (10.4)$$

where ψ is the helix angle of the thread. This can be rearranged to be

$$P = F\tan\psi \qquad (10.5)$$

Referring to the wedge in Figure 10.2, we can see that

$$\tan\psi = \frac{\ell}{\pi d} \qquad (10.6)$$

so that

$$P = \frac{F\ell}{\pi d} \qquad (10.7)$$

Since P is acting at a distance $d/2$ from the center of the screw, the torque acting on the screw, and therefore reacting on the disk, is

$$T = \left(\frac{F\ell}{\pi d}\right)\left(\frac{d}{2}\right) = \frac{F\ell}{2\pi} \tag{10.8}$$

The torque will give the disk an angular acceleration $\ddot{\theta}$ which has already been related to the translational acceleration of the free end $\ddot{x}$ in Equation 10.3. Equating the torque to the moment of inertia I of the disk multiplied by its angular acceleration leads to

$$I\ddot{\theta} = I\left(\frac{2\pi}{\ell}\right)\ddot{x} = \frac{F\ell}{2\pi} \tag{10.9}$$

Equation 10.9 can be rearranged to express the force acting on the moving connector F as a function of the acceleration of the threaded rod $\ddot{x}$.

$$F = \left(\frac{4\pi^2 I}{\ell^2}\right)\ddot{x} \tag{10.10}$$

The result is that the force is proportional to the acceleration and is therefore analogous to spring forces where we say $F = kx$ where k is the stiffness or damper forces where $F = c\dot{x}$ where c is the damping coefficient. We now have $F = b\ddot{x}$, where b is the inertia coefficient.

$$b = \frac{4\pi^2 I}{\ell^2} \tag{10.11}$$

Considering b as a design variable, we see that it varies linearly with disk moment of inertia and is inversely proportional to the square of the pitch of the screw. The relationship to the pitch is very important. To see this, consider the following example. Say we construct a device with a steel disk of radius 2 cm (0.02 m) and a thickness of 2 cm (0.02 m). With the density of steel ρ taken to be 7800 kg/m ^{3}, the mass can be worked out to be

$$m = \rho V = \rho \pi r^2 t = (7800)\pi(0.02)^2(0.02) = 0.196 \text{ kg}$$

The disk has a moment of inertia $I = mr^2/2$ so that b is

$$b = \frac{4\pi^2 I}{\ell^2} = \left(\frac{4\pi^2}{\ell^2}\right)\left(\frac{mr^2}{2}\right) = \frac{2\pi^2 mr^2}{\ell^2} = \frac{2\pi^2(0.196)(0.02)^2}{\ell^2} = \frac{0.00155}{\ell^2}$$

Table 10.1 shows how the inertia coefficient varies with pitch ℓ for a range of pitch (also called *lead*) values that is available in standard low-friction ball screws. The ratio of the inertia coefficient to the mass of the disk shows the dramatic increase in apparent mass of the system as the pitch decreases. At 2 mm pitch, the small rotating disk has the same effect as a adding a body with nearly 2000 times its mass to the system.

Figure 10.3 shows how the inerter is represented in a system. The force acting is

$$f = b(\ddot{x}_{out} - \ddot{x}_{in}) \tag{10.12}$$

where we now drop the requirement that one end be attached to ground as we have been considering to this point. Physically, if the inerter is attached to two bodies that have exactly the same acceleration (i.e. $\ddot{x}_{out} = \ddot{x}_{in}$), then there will be no angular acceleration of the disk and therefore no force. The angular acceleration of the disk will be proportional to the difference between the accelerations of the two connectors and thus the force will be that shown in Equation 10.12.

Table 10.1 Inertia coefficients.

ℓ	b	b/m
mm	kg	
10	15.5	79
5	62.0	316
2	388.0	1977

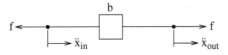

Figure 10.3 The inerter symbol.

10.2 The Inerter in the Equations of Motion

Figure 10.4 shows a single degree of freedom system with an inerter as well as the usual spring and damper. The free body diagram for this system is shown in Figure 10.5. A vertical force balance yields

$$-kx - c\dot{x} - b\ddot{x} = m\ddot{x} \tag{10.13}$$

which can be reorganized and written as

$$(m + b)\ddot{x} + c\dot{x} + kx = 0 \tag{10.14}$$

where it is clear that the inerter has simply added mass to the single degree of freedom system.

If we were to use Lagrange's Equation for the derivation, the inerter would add to the kinetic energy of the system. Considering the general element shown on Figure 10.3, the

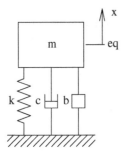

Figure 10.4 Single degree of freedom system with an inerter.

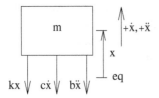

Figure 10.5 Free Body Diagram for the single degree of freedom system with an inerter.

kinetic energy supplied by the inerter would be

$$T = \frac{1}{2}b(\dot{x}_{out} - \dot{x}_{in})^2 \tag{10.15}$$

The kinetic energy for the single degree of freedom system is then

$$T = \frac{1}{2}m\dot{x}^2 + \frac{1}{2}b(\dot{x} - 0)^2 = \frac{1}{2}(m + b)\dot{x}^2 \tag{10.16}$$

and the inertial term in the equation of motion is

$$\frac{d}{dt}\left(\frac{\partial T}{\partial \dot{x}}\right) = (m + b)\ddot{x} \tag{10.17}$$

as it was shown to be in Equation 10.14.

We now turn our attention to a system with more than a single degree of freedom. Figure 10.6 shows a two degree of freedom system with two inerters. The kinetic energy for this system is

$$T = \frac{1}{2}m_1\dot{x}_1^2 + \frac{1}{2}m_2\dot{x}_2^2 + \frac{1}{2}b_1(\dot{x}_1 - 0)^2 + \frac{1}{2}b_2(\dot{x}_2 - \dot{x}_1)^2 \tag{10.18}$$

Rayleigh's dissipation function is

$$\mathcal{R} = \frac{1}{2}c_1(\dot{x}_1 - 0)^2 + \frac{1}{2}c_2(\dot{x}_2 - \dot{x}_1)^2 \tag{10.19}$$

and the potential energy

$$U = \frac{1}{2}k_1(x_1 - 0)^2 + \frac{1}{2}k_2(x_2 - x_1)^2 \tag{10.20}$$

Applying Lagrange's Equation gives the equations of motion as

$$
\begin{bmatrix} (m_1 + b_1 + b_2) & -b_2 \\ -b_2 & (m_2 + b_2) \end{bmatrix} \begin{Bmatrix} \ddot{x}_1 \\ \ddot{x}_2 \end{Bmatrix}
$$
$$
+ \begin{bmatrix} (c_1 + c_2) & -c_2 \\ -c_2 & c_2 \end{bmatrix} \begin{Bmatrix} \dot{x}_1 \\ \dot{x}_2 \end{Bmatrix}
$$
$$
+ \begin{bmatrix} (k_1 + k_2) & -k_2 \\ -k_2 & k_2 \end{bmatrix} \begin{Bmatrix} x_1 \\ x_2 \end{Bmatrix} = \begin{Bmatrix} 0 \\ 0 \end{Bmatrix} \tag{10.21}
$$

Figure 10.6 Two degree of freedom system with inerters.

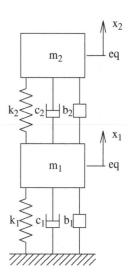

The inerter connected to ground (b_1) has simply added mass to m_1 just as it did in the single degree of freedom case. On the other hand, the inerter connecting the two masses (b_2) has significantly changed the equations of motion. It adds mass to both m_1 and m_2 as can be seen in the diagonal elements in the mass matrix. It also introduces inertial coupling as can be seen by the non-zero off-diagonal terms. Before the inerter was introduced, inertial coupling only came into play when the analysis was done using relative coordinates. x_1 and x_2 in this example are absolute coordinates and would never have been coupled in the mass matrix previously.

10.3 An Examination of the Effect of an Inerter on System Response

While it is impossible to consider all possible configurations of systems including inerters, we can look at how an inerter changes the response of a simplified system with two degrees of freedom to gain some understanding of its effect. The system in Figure 10.7 has two equal masses, no damping, two identical springs, and a single inerter with b equal to some fraction p of the mass of one or the other bodies (i.e. $b = pm$ where p is dimensionless).

Substituting into Equation 10.21 gives the equations of motion as

$$\begin{bmatrix} (1+p)m & -pm \\ -pm & (1+p)m \end{bmatrix} \begin{Bmatrix} \ddot{x}_1 \\ \ddot{x}_2 \end{Bmatrix} + \begin{bmatrix} 2k & -k \\ -k & k \end{bmatrix} \begin{Bmatrix} x_1 \\ x_2 \end{Bmatrix} = \begin{Bmatrix} 0 \\ 0 \end{Bmatrix} \tag{10.22}$$

Assuming harmonic solutions at frequency ω with amplitudes X_1 and X_2 gives

$$\begin{bmatrix} -\omega^2(1+p)m + 2k & \omega^2 pm - k \\ \omega^2 pm - k & -\omega^2(1+p)m + k \end{bmatrix} \begin{Bmatrix} X_1 \\ X_2 \end{Bmatrix} = \begin{Bmatrix} 0 \\ 0 \end{Bmatrix} \tag{10.23}$$

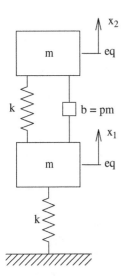

Figure 10.7 Simplified two degree of freedom system with an inerter.

then divide by k everywhere and define $\lambda = \omega^2 m/k$ to get

$$\begin{bmatrix} -\lambda(1+p)+2 & \lambda p - 1 \\ \lambda p - 1 & -\lambda(1+p)+1 \end{bmatrix} \begin{Bmatrix} X_1 \\ X_2 \end{Bmatrix} = \begin{Bmatrix} 0 \\ 0 \end{Bmatrix} \tag{10.24}$$

The characteristic equation is

$$\lambda^2 (1+2p) - \lambda(3+p) + 1 = 0 \tag{10.25}$$

with roots

$$\lambda = \frac{(3+p) \pm \sqrt{p^2 - 2p + 5}}{2(1+2p)} \tag{10.26}$$

where λ reflects the ratio of a natural frequency of the system to the natural frequency of a single degree of freedom system with mass m and stiffness k. That is

$$\lambda = \omega^2 m/k = \left(\frac{\omega}{\omega_n}\right)^2 \text{ where } \omega_n = \sqrt{k/m} \tag{10.27}$$

The factor p can theoretically take on any value. We will consider three values of p to see how the behavior of the system differs. $p = 0$ will give the natural frequencies and mode shapes for the system without the inerter. The case where $p = 1$ is interesting and will be considered. The case where p is very large will give the limiting values of natural frequencies and mode shapes.

10.3.1 The Baseline Case – $p = 0$

For $p = 0$, Equation 10.24 becomes

$$\begin{bmatrix} -\lambda + 2 & -1 \\ -1 & -\lambda + 1 \end{bmatrix} \begin{Bmatrix} X_1 \\ X_2 \end{Bmatrix} = \begin{Bmatrix} 0 \\ 0 \end{Bmatrix} \tag{10.28}$$

with characteristic equation

$$\lambda^2 - 3\lambda + 1 = 0 \tag{10.29}$$

The roots are

$$\lambda_1 = \frac{3 - \sqrt{5}}{2} \text{ and } \lambda_2 = \frac{3 + \sqrt{5}}{2}$$

corresponding to natural frequencies

$$\omega_1 = 0.618\sqrt{\frac{k}{m}} \text{ and } \omega_2 = 1.618\sqrt{\frac{k}{m}}$$

The corresponding mode shapes are

$$\begin{Bmatrix} X_1 \\ X_2 \end{Bmatrix}_1 = \begin{Bmatrix} 1 \\ 1.618 \end{Bmatrix} \text{ and } \begin{Bmatrix} X_1 \\ X_2 \end{Bmatrix}_2 = \begin{Bmatrix} 1 \\ -0.618 \end{Bmatrix}$$

10.3.2 The Case Where the Inerter Adds Mass Equal to the Block's Mass – $p = 1$

Equation 10.24 becomes

$$
\begin{bmatrix} -2\lambda + 2 & \lambda - 1 \\ \lambda - 1 & -2\lambda + 1 \end{bmatrix} \begin{Bmatrix} X_1 \\ X_2 \end{Bmatrix} = \begin{Bmatrix} 0 \\ 0 \end{Bmatrix}
\tag{10.30}
$$

with roots found using Equation 10.26

$$
\lambda_1 = \frac{1}{3} \text{ and } \lambda_2 = 1
$$

The mode shapes are

$$
\begin{Bmatrix} X_1 \\ X_2 \end{Bmatrix}_1 = \begin{Bmatrix} 1 \\ 2 \end{Bmatrix} \text{ and } \begin{Bmatrix} X_1 \\ X_2 \end{Bmatrix}_2 = \begin{Bmatrix} 1 \\ 0 \end{Bmatrix}
$$

The mode shape corresponding to the higher natural frequency (i.e. $\lambda_2 = 1$ so $\omega_2 = \sqrt{k/m}$) is interesting because it has no motion of the of the upper block (i.e. $X_2 = 0$). We can think of the upper block acting as ground since it isn't moving. As a result, the inerter simply doubles the mass of the lower block leading to a mass of $2m$ supported by two springs each having stiffness k. The resulting single degree of freedom system then has a natural frequency

$$
\omega_n = \sqrt{\frac{2k}{2m}} = \sqrt{\frac{k}{m}}
$$

as indicated by $\lambda_2 = 1$.

10.3.3 The Case Where p is Very Large

Here, we are considering cases where the inerter adds a great deal of mass to the system. In considering the limits for large values of p, we will have to deal with the square root term in Equation 10.26

$$
\sqrt{p^2 - 2p + 5}
$$

There is a tendency to say that "for large p the p^2 term dominates so this simplifies to $\sqrt{p^2} \approx p$", but that turns out to be inadequate in this case. In fact, the Binomial Theorem[2] has to be invoked to get the correct limits. First, write

$$
\sqrt{p^2 - 2p + 5} = \sqrt{(p-1)^2 + 4}
$$

then, use the first two terms from the Binomial Theorem[2] to approximate the expression as

$$
\sqrt{(p-1)^2 + 4} \approx (p-1) + \frac{1}{2} \frac{4}{(p-1)} \approx (p-1) + \frac{2}{(p-1)}
$$

---

2 The Binomial Theorem is an algebraic expansion of the sum of two terms multiplied together n times. It can be expressed as

$$
(a+b)^n = a^n + na^{n-1}b + \frac{n(n-1)}{2}a^{n-2}b^2 + \frac{n(n-1)(n-2)}{6}a^{n-3}b^3 + \cdots
$$

In this case, the expansion is of

$$
[(p-1)^2 + 4]^{1/2}
$$

so $a = (p-1)^2$, $b = 4$, and $n = 1/2$

Equation 10.26 then gives

$$\lambda = \frac{(3+p) \pm \left[(p-1) + \frac{2}{(p-1)}\right]}{2(1+2p)}$$ (10.31)

For large values of p, $2/(p-1)$ becomes negligible and the result is

$$\lambda_1 = \frac{(3+p) - (p-1)}{2+4p} \approx \frac{4}{4p} = \frac{1}{p}$$ (10.32)

and

$$\lambda_2 = \frac{(3+p) + (p-1)}{2+4p} \approx \frac{2p}{4p} = \frac{1}{2}$$ (10.33)

where we note that the Binomial Theorem term was not important to the eigenvalues but we will see that it is necessary for finding the mode shapes.

To find the mode shapes, we refer to Equation 10.24, which gives two equations relating the amplitudes X_1 and X_2. The first equation is

$$[-\lambda(1+p) + 2]X_1 + (\lambda p - 1)X_2 = 0$$

from which

$$\frac{X_2}{X_1} = \frac{[\lambda(1+p) - 2]}{(\lambda p - 1)}$$ (10.34)

The second equation is

$$(\lambda p - 1)X_1 + [-\lambda(1+p) + 1]X_2 = 0$$

giving

$$\frac{X_2}{X_1} = \frac{(\lambda p - 1)}{[\lambda(1+p) - 1]}$$ (10.35)

Consider finding the mode shape for $\lambda_1 = 1/p$. If we use Equation 10.34, we get

$$\frac{X_2}{X_1} = \frac{\left[\frac{1}{p}(1+p) - 2\right]}{\left(\frac{1}{p}p - 1\right)} = \frac{\left(\frac{1}{p} + 1 - 2\right)}{(1-1)} = \frac{\left(\frac{1}{p} - 1\right)}{0} = -\infty$$

and, if we use Equation 10.35, we get

$$\frac{X_2}{X_1} = \frac{\left(\frac{1}{p}p - 1\right)}{\left[\frac{1}{p}(1+p) - 1\right]} = \frac{(1-1)}{\left(\frac{1}{p} + 1 - 1\right)} = \frac{0}{\frac{1}{p}} = 0$$

There is clearly a problem here since X_2/X_1 cannot be both infinite and zero at the same time.

Rather than using $1/p$ as an approximation for λ_1, we add the Binomial Theorem term and use

$$\lambda_1 = \frac{(3+p) - \left[(p-1) + \frac{2}{(p-1)}\right]}{2(1+2p)}$$

which can be simplified to

$$\lambda_1 = \frac{(2p - 3)}{(p - 1)(1 + 2p)}$$

Substituting this into Equation 10.34, simplifying, and evaluating for large p gives

$$\frac{X_2}{X_1} = \frac{\frac{(2p-3)(1+p)}{(p-1)(1+2p)} - 2}{\frac{(2p-3)p}{(p-1)(1+2p)} - 1} = \frac{-2p^2 + p - 1}{-2p + 1} \approx p$$

Doing the same thing with Equation 10.35 yields

$$\frac{X_2}{X_1} = \frac{\frac{(2p-3)p}{(p-1)(1+2p)} - 1}{\frac{(2p-3)(1+p)}{(p-1)(1+2p)} - 1} = \frac{-2p + 1}{-2} \approx p$$

The result is that as p gets very large $\lambda_1 = (\omega/\omega_n)^2$ approaches $1/p$ which in turn approaches zero. The natural frequency is therefore very low and the mode shape has X_2/X_1 growing in proportion to p so that motions will be predominantly of the upper block with the lower block becoming essentially stationary for very large values of p.

For $\lambda_2 = 1/2$, both Equations 10.34 and 10.35 give the same result. Equation 10.34 gives, for large p,

$$\frac{X_2}{X_1} = \frac{\left[\frac{1}{2}(1 + p) - 2\right]}{\left(\frac{p}{2} - 1\right)} = \frac{\left(\frac{1}{2} + \frac{p}{2} - 2\right)}{\left(\frac{p}{2} - 1\right)} = \frac{\frac{p}{2} - \frac{3}{2}}{\frac{p}{2} - 1} \approx 1$$

and Equation 10.35 gives

$$\frac{X_2}{X_1} = \frac{\left(\frac{p}{2} - 1\right)}{\left[\frac{1}{2}(1 + p) - 1\right]} = \frac{\frac{p}{2} - 1}{\frac{p}{2} - \frac{1}{2}} \approx 1$$

Since X_1 and X_2 have no relative motion in this mode, the inerter will have no effect so the mode has both blocks (total mass $= 2m$) supported by the single spring connecting the bottom block to the ground and the natural frequency must then be $\omega = \sqrt{k/2m}$ which corresponds to $\lambda_2 = 1/2$.

10.4 The Inerter as a Vibration Absorber

In this section we look at the common case of vibrations induced by harmonic base motions. Previously we had considered the case of the "tuned vibration absorber" where a sacrificial mass and spring (the "vibration absorber") were attached to a system experiencing forced vibrations and the natural frequency of the single degree of freedom system made by the mass and spring was tuned to coincide with the forcing frequency. As a result, the absorber responded to the forcing frequency and the main system stopped vibrating. Here we consider controlling the vibration with an inerter.

Figure 10.8 shows a single degree of freedom system disturbed by harmonic ground motion $y(t) = Y \sin \omega t$. ω is the forcing frequency and, if it is near to the natural frequency,

Figure 10.8 A single degree of freedom system with harmonic ground motion.

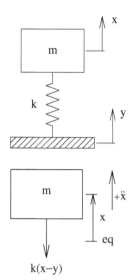

Figure 10.9 Free body diagram for the single degree of freedom system with harmonic ground motion.

$\omega_n = \sqrt{k/m}$, the motions of the mass will be large. In the steady state, $x(t)$ will be a harmonic function having the same frequency as $y(t)$. From the free body diagram shown in Figure 10.9, the equation of motion for the system is

$$+\uparrow \sum F : -k(x-y) = m\ddot{x} \text{ or } m\ddot{x} + kx = ky \qquad (10.36)$$

Substituting $y(t) = Y \sin \omega t$, $x(t) = X \sin \omega t$, and $\ddot{x}(t) = -\omega^2 X \sin \omega t$ gives

$$(-\omega^2 m + k)X \sin \omega t = kY \sin \omega t \qquad (10.37)$$

Equation 10.37 allows the ratio of the output amplitude X to the input or disturbance amplitude Y to be written as

$$\frac{X}{Y} = \frac{k}{(-\omega^2 m + k)} \qquad (10.38)$$

which can be divided throughout by k to yield

$$\frac{X}{Y} = \frac{1}{1 - \left(\frac{\omega}{\omega_n}\right)^2} \qquad (10.39)$$

where it is clear that the denominator decreases as the forcing frequency approaches the natural frequency and the amplitude of motion of the mass will increase.

Consider the same system but with an inerter added as shown in Figure 10.10[3]. Figure 10.11 shows the free body diagram for the modified system. The inerter coefficient is taken as a proportion of the mass m (i.e. $b = pm$ where p is a dimensionless design parameter).

The force balance is now

$$+\uparrow \sum F : -k(x-y) - pm(\ddot{x} - \ddot{y}) = m\ddot{x} \qquad (10.40)$$

3 In Section 10.1 the use of an inerter connected to ground was said to add mass. That was the case because the ground wasn't moving so the inerter was sensing absolute acceleration. Here the ground has an acceleration so the inerter responds to relative acceleration.

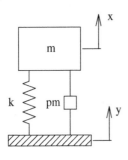

Figure 10.10 A single degree of freedom system with an inerter and harmonic ground motion.

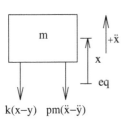

Figure 10.11 Free body diagram for the single degree of freedom system with an inerter and harmonic ground motion.

resulting in the equation of motion

$$(1 + p)m\ddot{x} + kx = pm\ddot{y} + ky \tag{10.41}$$

The extra mass that we knew would appear shows up on the left-hand side of Equation 10.41 as the pm term and, more importantly, there is an extra forcing function $pm\ddot{y}$ that has been introduced on the right-hand side.

Assuming harmonic input and output at the forcing frequency ω as before gives

$$[-\omega^2 (1 + p)m + k]X \sin \omega t = [-\omega^2 pm + k]Y \sin \omega t \tag{10.42}$$

and the ratio X/Y becomes

$$\frac{X}{Y} = \frac{k - \omega^2 pm}{k - \omega^2 (1 + p)m} \tag{10.43}$$

which can be divided throughout by k to yield[4]

$$\frac{X}{Y} = \frac{1 - p\left(\frac{\omega}{\omega_n}\right)^2}{1 - (1 + p)\left(\frac{\omega}{\omega_n}\right)^2} \tag{10.44}$$

The result presented in Equation 10.44 shows that the amplitude ratio X/Y, and therefore the response of the system, can be made zero through an appropriate choice of p. The numerator, and therefore X, goes to zero when

$$1 - p\left(\frac{\omega}{\omega_n}\right)^2 = 0 \tag{10.45}$$

---

4 The definition of $\omega_n = \sqrt{k/m}$ has been retained here even though it is no longer the natural frequency of the system because the inerter has changed the mass. The natural frequency is now $\sqrt{k/(1+p)m}$ which leads to an unwieldy expression for X/Y.

This requires that

$$p = \left(\frac{\omega_n}{\omega}\right)^2 = \frac{k}{m\omega^2} \qquad (10.46)$$

The denominator of Equation 10.44 is non-zero for this value of p so X will indeed be zero and the vibration absorber has been realized.

There are two different effects that the inerter has on the system. The first is the change in natural frequency that is discussed in Footnote 4. The second, and more important, effect is the added forcing function resulting from the acceleration of the ground. This was the $pm\ddot{y}$ term that was remarked upon just after Equation 10.41. Equation 10.42 resulted from the assumption of harmonic motion and the forcing function became

$$[-\omega^2 pm + k]Y \sin \omega t \qquad (10.47)$$

where it is clear that the force passing through the spring and the force passing through the inerter have opposite signs and setting p equal to $k/(m\omega^2)$ as suggested in Equation 10.46 actually causes the two forcing function terms to become equal and opposite so that there is no net force transmitted to the system and it stops moving.

Exercises

10.1 For the system shown in Figure 10.10, plot the amplitude ratio (Equation 10.44) as a function of the frequency ratio. Use values of $p = 0$, $p = 0.1$, $p = 1$, and $p = 10$ and let the frequency ratio vary from 0 to 4. Comment on the results.

10.2 Write an expression for the force transmissibility, F_T/kY, for the system shown in Figure 10.10. Plot the force transmissibility as a function of the frequency ratio. Use values of $p = 0$, $p = 0.1$, $p = 1$, and $p = 10$ and let the frequency ratio vary from 0 to 4. Comment on how these results could be used if you were using an inerter in one of your designs.

10.3 Consider a system with a rotating unbalance such as we covered in Section 5.4. Derive an expression for the Amplitude Ratio defined in Equation 5.74 for the case where the damper has been replaced by an inerter with $b = pm$.

11

Analysis of Experimental Data

Up to now, we have been concentrating on the theoretical simulation of systems in order to obtain information useful in their design. The theoretical study of dynamic systems and the ability to understand the physical behavior of the systems that the theory imparts are immensely important to practicing engineers. The theory is what leads to the initial system design through simulation and the theory is what gives the ability to understand and correct the behavior of the prototype systems when they are tested.

Equally as important to engineers is the analysis of experimentally derived response data for systems. The validation of the models used in the design phase relies on being able to record and analyze test data. Such analysis is most often done in the frequency domain and that is what we will concentrate on here.

11.1 Typical Test Data

Dynamic measurements are made by sensors of various kinds and are most often recorded in digital form[1]. Figure 11.1 shows a typical measured dynamic variable, $x(t)$, plotted as a function of time. Measured variables very often appear to be random because, even after looking at the signal over the entire plot, there is no way to predict where the next point will be.

Characterization of the time domain signal is unrewarding. The mean value of the signal can be found but is not representative of anything but the "average" value of $x(t)$. It contains no information with respect to peak values for instance. Given that $x(t)$ is known in the range $0 \leq t \leq T$, the mean value is defined as

$$\bar{x} = \frac{1}{T} \int_0^T x(t) \, dt \tag{11.1}$$

---

[1] The analog measurements that used to be recorded on analog devices such as multi-channel FM tape recorders have been almost completely superseded by in-field A/D converters and digital recording devices. The sampling rate of the A/D converters must be chosen very carefully to satisfy the requirements of the data analysis that will be done later. With analog devices it was possible to re-play the experiments if you got the sampling rate wrong. Now you are required to get it right or repeat the, sometimes very costly, experiments.

Introduction to Mechanical Vibrations, First Edition. Ronald J. Anderson.
© 2020 John Wiley & Sons Ltd. Published 2020 by John Wiley & Sons Ltd.
Companion website: www.wiley.com/go/anderson/introduction-to-vibrations

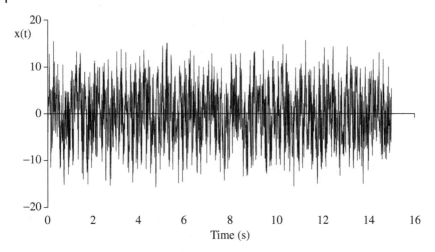

Figure 11.1 A measured variable $x(t)$ plotted versus time.

A characteristic of the signal that can be calculated and that has meaning is the total *mean-square value* of the signal, $\overline{x}_T^2$. The total mean-square value of the signal is defined as the mean value of another signal generated by squaring the original signal and then finding its mean value over the period $0 \le t \le T$ using Equation 11.2. Figure 11.2 shows the squared signal and the mean-square value $\overline{x}_T^2 = 43$. Characterizing the signal by this single scalar value seems less than satisfying but that is all that can be done in the time domain. However, the total mean-square value will be useful to us when we discuss scaling

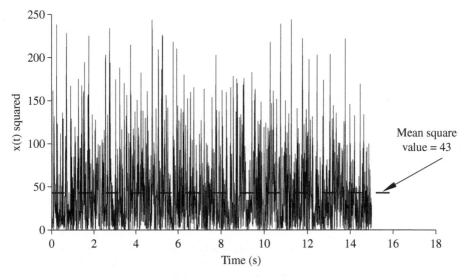

Figure 11.2 The square of $x(t)$ plotted versus time.

of results in the frequency domain in Section 11.8.

$$\bar{x}_T^2 = \frac{1}{T} \int_0^T x^2(t)dt \tag{11.2}$$

11.2   Transforming to the Frequency Domain – The CFT

The most common approach to analyzing experimental signals begins with a transformation that takes the measured data into the frequency domain. Readers may be familiar with the terms *FFT* and *Fast Fourier Transform* that describe a technique that is widely used to accomplish the transformation but may not familiar with the details of the transformation. The mathematical development of the Fourier Transform equations is most easily done using complex exponentials and that is the approach used in modern reference books. Unfortunately, the complex exponential approach, while being mathematically elegant, leaves the reader without any physical feeling for what lies behind the equations. The goal here is to present the techniques in a manner that leaves the reader with an understanding of the techniques and a physical feeling for what they represent.

It is best to think of the transformation as a curve fitting exercise. That is, we have a continuous time-domain function $x(t)$ and we wish to approximate it in the range $0 \le t \le T$ by using a series of harmonic functions. We choose a base frequency f_0 that allows one complete cycle in the range and then add higher-frequency components that are multiples of f_0. Clearly, f_0 in cycles per second will be such that there is one cycle in T seconds so that $f_0 = 1/T$ cycles per second or Hz. The base frequency in radians per second will then be $\omega_0 = 2\pi/T$.

Let the approximating function $x_A(t)$ be expressed in terms of $(2N + 1)$ undetermined coefficients a_0, a_n, and b_n as follows

$$x_A(t) = a_0 + \sum_{n=1}^{N} a_n \cos(n\omega_0 t) + \sum_{n=1}^{N} b_n \sin(n\omega_0 t) \tag{11.3}$$

While there are several techniques that could be used to find the "best" values of the coefficients, we will use the traditional Least Squares[2] approach because it is commonly used in fitting curves to experimental data. We first define an objective function J which is a measure of the difference between the experimental data and the approximating function. Since the difference $[x(t) - x_A(t)]$ can be either positive or negative, we can't use it to establish the goodness of fit so we work with the square of the difference instead and define:

$$J = \frac{1}{2} \int_0^T [x(t) - x_A(t)]^2 dt \tag{11.4}$$

We choose the coefficients such that they minimize the objective function, thereby minimizing the difference between the experimental curve $x(t)$ and the approximating function $x_A(t)$. The conditions for minimizing J are that the partial derivatives of J with respect to all

---

2 Appendix A has a description of Least Squares curve fitting for those who would like to review the method before proceeding with the remainder of this chapter.

of the undetermined coefficients must simultaneously be equal to zero. J can be expanded into[3].

$$J = \frac{1}{2} \int_0^T \left[x(t) - a_0 - \sum_{n=1}^N a_n \cos(n\omega_0 t) - \sum_{n=1}^N b_n \sin(n\omega_0 t) \right]^2 dt \qquad (11.5)$$

The partial derivative with respect to a_0 is

$$\frac{\partial J}{\partial a_0} = \int_0^T \left[-x(t) + a_0 + \sum_{n=1}^N a_n \cos(n\omega_0 t) \right.$$
$$\left. + \sum_{n=1}^N b_n \sin(n\omega_0 t) \right] dt = 0 \qquad (11.6)$$

For the remaining coefficients,

$$\frac{\partial J}{\partial a_p} = \int_0^T \left[-x(t) + a_0 + \sum_{n=1}^N a_n \cos(n\omega_0 t) \right.$$
$$\left. + \sum_{n=1}^N b_n \sin(n\omega_0 t) \right] \cos(p\omega_0 t) dt = 0, \ p = 1, N \qquad (11.7)$$

and

$$\frac{\partial J}{\partial b_p} = \int_0^T \left[-x(t) + a_0 + \sum_{n=1}^N a_n \cos(n\omega_0 t) \right.$$
$$\left. + \sum_{n=1}^N b_n \sin(n\omega_0 t) \right] \sin(p\omega_0 t) dt = 0, \ p = 1, N \qquad (11.8)$$

Consider first satisfying the condition on a_0 from Equation 11.6. This can be written as

$$-\int_0^T x(t)dt + a_0 \int_0^T dt + \sum_{n=1}^N a_n \left[\int_0^T \cos(n\omega_0 t)dt \right]$$
$$+ \sum_{n=1}^N b_n \left[\int_0^T \sin(n\omega_0 t)dt \right] = 0 \qquad (11.9)$$

Three of the four integrals in Equation 11.9 can be directly evaluated, yielding

$$\int_0^T dt = T; \quad \int_0^T \cos(n\omega_0 t)dt = 0 \ ; \quad \int_0^T \sin(n\omega_0 t)dt = 0 \qquad (11.10)$$

The integrals of the trigonometric functions give the area under them as time goes from zero to T. Since they have been defined as multiples of a single wavelength spanning $0 \le t \le T$, they will all complete an integer number of cycles and the area under them will be zero. A mathematical proof is left as an exercise.

---

3 You may wonder why a_0 is included and b_0 is not. In fact, a more elegant way of writing the series is

$$x_A(t) = \sum_{n=0}^N a_n \cos(n\omega_0 t) + \sum_{n=0}^N b_n \sin(n\omega_0 t)$$

but it is clear that $\sin(n\omega_0 t) = 0$ when $n = 0$ so the form shown in Equation 11.3 is universal.

The result is that a_0 is

$$a_0 = \frac{1}{T} \int_0^T x(t)dt \tag{11.11}$$

Comparing this to Equation 11.1, we see that a_0 is the mean value of $x(t)$. Examination of the approximating function in Equation 11.3 makes it clear that this must be the case. The function has one constant term (a_0) and superimposes harmonics on top of it. The constant value can only be the mean value.

We now turn to the conditions on the other coefficients as specified in Equations 11.7 and 11.8. These involve integrals of the form

$$\int_0^T \cos(n\omega_0 t) \sin(p\omega_0 t)dt \tag{11.12}$$

$$\int_0^T \sin(n\omega_0 t) \sin(p\omega_0 t)dt \tag{11.13}$$

$$\int_0^T \cos(n\omega_0 t) \cos(p\omega_0 t)dt \tag{11.14}$$

and

$$\int_0^T \sin(n\omega_0 t) \cos(p\omega_0 t)dt \tag{11.15}$$

all of which can be shown to be zero so long as $n \neq p$ (see Exercise 11.2). This is a result of the functions 1, $\cos pt$, and $\sin pt$ being *orthogonal* over the interval $[0 \leq pt \leq 2\pi]$. This concept of orthogonality will reappear when we consider Discrete Fourier Transforms in Section 11.4.

In the case where $p = n$ Equations 11.7 and 11.8 become

$$-\int_0^T x(t)\cos(n\omega_0 t)dt + a_0 \int_0^T \cos(n\omega_0 t)dt + a_n \int_0^T \cos^2(n\omega_0 t)dt$$

$$+ b_n \int_0^T \sin(n\omega_0 t)\cos(p\omega_0 t)dt = 0 \tag{11.16}$$

and

$$-\int_0^T x(t)\sin(p\omega_0 t)dt + a_0 \int_0^T \sin(p\omega_0 t)dt + a_n \int_0^T \cos(n\omega_0 t)\sin(p\omega_0 t)dt$$

$$+ b_n \int_0^T \sin^2(n\omega_0 t)dt = 0 \tag{11.17}$$

Several of the integrals in Equations 11.16 and 11.17 have already been shown to be zero. In addition, it can easily be shown that

$$\int_0^T \cos^2(n\omega_0 t)dt = \int_0^T \sin^2(n\omega_0 t)dt = \frac{T}{2} \tag{11.18}$$

and, as a result, we find that the Least Squares curve fit to the experimental data is

$$x_A(t) = a_0 + \sum_{n=1}^N a_n \cos(n\omega_0 t) + \sum_{n=1}^N b_n \sin(n\omega_0 t) \tag{11.19}$$

where

$$a_0 = \frac{1}{T} \int_0^T x(t)dt \tag{11.20}$$

$$a_n = \frac{2}{T} \int_0^T x(t)\cos(n\omega_0 t)dt, \quad n = 1, N \tag{11.21}$$

$$b_n = \frac{2}{T} \int_0^T x(t)\sin(n\omega_0 t)dt, \quad n = 1, N \tag{11.22}$$

If we now let N increase so that we minimize the objective function at more and more points we see that as N approaches infinity the approximation becomes equal to the function $x(t)$ so that we can write

$$x(t) = a_0 + \sum_{n=1}^{\infty} a_n \cos(n\omega_0 t) + \sum_{n=1}^{\infty} b_n \sin(n\omega_0 t) \tag{11.23}$$

where

$$a_0 = \frac{1}{T} \int_0^T x(t)dt \tag{11.24}$$

$$a_n = \frac{2}{T} \int_0^T x(t)\cos(n\omega_0 t)dt, \quad n = 1, \infty \tag{11.25}$$

$$b_n = \frac{2}{T} \int_0^T x(t)\sin(n\omega_0 t)dt, \quad n = 1, \infty \tag{11.26}$$

The transformation given by Equations 11.23 through 11.26 is the well known Continuous Fourier Transform (CFT)[4].

Notice that we have considered a continuous function $x(t)$ that was defined over the the interval $0 \leq t \leq T$. The Fourier Series representation of this function is based on this interval even though it can be evaluated for any time t. There is an implicit assumption that the function being analyzed is periodic with a period equal to T. This fact will become important to us as we proceed.

11.3 Transforming to the Frequency Domain – The DFT

We have shown that a periodic, continuous function can be exactly represented by a curve fit using an infinite series of harmonic functions, with a base frequency derived from the period of the function and multiples of that frequency. Clearly, practicing engineers will never have to deal with functions like this. Measurements we take are not continuous but are sampled and stored at some sampling rate that gives an equal time between samples, Δt. The measured values are not periodic but we have to finish sampling at some time so

---

4 Jean Baptiste Joseph Fourier (1768–1830), a French mathematician who was heavily involved in the politics of the French revolution and associated with Napoleon, is best remembered for his work on the propagation of heat in solid bodies and for his expansions of functions as trigonometric series that we now call Fourier series.

we get data over the range $0 \le t \le T$ where T is the time at which we stop sampling. As a result we get a finite number of points to analyze. Consider the curve fitting process again but without a continuous function to fit this time.

Let there be $2N$ measured data points stored in a vector $x(t)$. Let the sampling rate be $f_s = 1/\Delta t$ where Δt is the constant time between samples. Define the base frequency f_0 in Hz as

$$f_0 = \frac{1}{2N\Delta t} \tag{11.27}$$

giving a base frequency ω_0 in rad/s

$$\omega_0 = \frac{2\pi}{2N\Delta t} = \frac{\pi}{N\Delta t} \tag{11.28}$$

Use a series of harmonics to fit the measured data. The series is

$$x(t) = \sum_{n=0}^{N} [a_n \cos(n\omega_0 t) + b_n \sin(n\omega_0 t)] \tag{11.29}$$

Since we have $2N$ data points, we can use them to find exactly $2N$ coefficients, a_n and b_n, in Equation 11.29. As it stands, Equation 11.29 has $(2N + 2)$ undetermined coefficients. We can rewrite the series with its first and last terms extracted to see which two coefficients are unnecessary. We do this for some time $t = m\Delta t$ where $0 \le m \le (2N - 1)$. Substituting Equation 11.28 into Equation 11.29 and extracting the first and last terms gives $2N$ simultaneous equations of the form

$$x_m = x(m\Delta t) = a_0 \cos(0) + b_0 \sin(0)$$

$$+ \sum_{n=1}^{N-1} \left[a_n \cos\left(\frac{n\pi}{N}m\right) + b_n \sin\left(\frac{n\pi}{N}m\right) \right]$$

$$+ a_N \cos(\pi m) + b_N \sin(\pi m), \quad 0 \le m \le (2N - 1) \tag{11.30}$$

where we note that $\cos(0) = 1$ and $\sin(0) = \sin(\pi m) = 0$ so that the coefficients b_0 and b_N do not contribute to the curve fit and can be removed[5]. This leaves us with exactly $2N$ undetermined coefficients and $2N$ data points so we can write $2N$ simultaneous equations of the form

$$x_m = a_0 + \sum_{n=1}^{N-1} \left[a_n \cos\left(\frac{n\pi}{N}m\right) + b_n \sin\left(\frac{n\pi}{N}m\right) \right]$$

$$+ a_N \cos(\pi m), \quad 0 \le m \le (2N - 1) \tag{11.31}$$

The equations can be assembled into the standard matrix form for linear, algebraic equations with a known coefficient matrix, $[C]$, multiplied a vector of unknowns,

---

5 We are treating the case where we have an even number of samples. The case for an odd number of samples, $(2N + 1)$, can be handled in a similar way with the small difference that an extra coefficient, b_N, is retained.

$\{a_0 \dots a_N \, b_1 \dots b_{N-1}\}^T$, set equal to a vector of known values $\{x_0 \dots x_{2N-1}\}^T$

$$
[C] \begin{Bmatrix} a_0 \\ \vdots \\ a_N \\ b_1 \\ \vdots \\ b_{N-1} \end{Bmatrix} = \begin{Bmatrix} x_0 \\ \vdots \\ x_{2N-1} \end{Bmatrix} \tag{11.32}
$$

where the 2N by 2N coefficient matrix, [C], has terms c_{ij}, for the a_n (left side of the matrix)

$$
c_{ij} = \cos\left[\frac{(i-1)(j-1)\pi}{N}\right] ; i = 1, 2N, j = 1, N+1 \tag{11.33}
$$

and, for the b_n (right side of the matrix)

$$
c_{ij} = \sin\left[\frac{(i-1)(j-N-1)\pi}{N}\right] ; i = 1, 2N, j = N+2, 2N \tag{11.34}
$$

Solving for the unknown coefficients using Equation 11.32 is a simple matter of using a standard elimination technique such as Gaussian Elimination. The computational effort required to do so is approximately proportional to N^3 where "computational effort" is measured in terms of the number of multiplications and divisions required to implement a solution on a digital computer.

11.4 Transforming to the Frequency Domain – A Faster DFT

Here, we return to the concept of Least Squares curve fitting that we used in Section 11.2 but now it is applied to the finite number of data points we discussed in Section 11.3. We start with the series from Equation 11.31 rewritten so that a_0 and a_N are taken into the summation

$$
x_m = \sum_{n=0}^{N} a_n \cos\left(\frac{n\pi}{N}m\right) + \sum_{n=1}^{N-1} b_n \sin\left(\frac{n\pi}{N}m\right), 0 \le m \le (2N-1) \tag{11.35}
$$

and construct an objective function, J, as

$$
J = \frac{1}{2}\sum_{m=0}^{2N-1} \left[x_m - \sum_{n=0}^{N} a_n \cos\left(\frac{n\pi}{N}m\right) - \sum_{n=1}^{N-1} b_n \sin\left(\frac{n\pi}{N}m\right) \right]^2 \tag{11.36}
$$

As before, we set the partial derivatives of J with respect to each of the coefficients to zero. That is,

$$
\frac{\partial J}{\partial a_p} = -\sum_{m=0}^{2N-1} x_m \cos\left(\frac{p\pi}{N}m\right) + \sum_{m=0}^{2N-1}\left[\sum_{n=0}^{N} a_n \cos\left(\frac{n\pi}{N}m\right)\cos\left(\frac{p\pi}{N}m\right) \right.
$$
$$
\left. + \sum_{n=1}^{N-1} b_n \sin\left(\frac{n\pi}{N}m\right)\cos\left(\frac{p\pi}{N}m\right) \right] = 0 , p = 0, N \tag{11.37}
$$

and

$$\frac{\partial J}{\partial b_p} = -\sum_{m=0}^{2N-1} x_m \sin\left(\frac{p\pi}{N}m\right) + \sum_{m=0}^{2N-1}\left[\sum_{n=0}^{N} a_n \cos\left(\frac{n\pi}{N}m\right)\sin\left(\frac{p\pi}{N}m\right)\right.$$

$$\left. + \sum_{n=1}^{N-1} b_n \sin\left(\frac{n\pi}{N}m\right)\sin\left(\frac{p\pi}{N}m\right)\right] = 0 \, , p = 1, N-1 \tag{11.38}$$

We used orthogonality in Section 11.2 to show that the vast majority of the integrals involving products of harmonic functions were equal to zero. Orthogonality also holds in the case of summations. The rules are as follows

$$\sum_{m=0}^{2N-1} \sin\left(\frac{n\pi}{N}m\right)\sin\left(\frac{p\pi}{N}m\right) = \begin{cases} 0 \text{ if } p \neq n \\ N \text{ if } p = n = 0 \end{cases} \tag{11.39}$$

$$\sum_{m=0}^{2N-1} \sin\left(\frac{n\pi}{N}m\right)\cos\left(\frac{p\pi}{N}m\right) = 0 \tag{11.40}$$

$$\sum_{m=0}^{2N-1} \cos\left(\frac{n\pi}{N}m\right)\cos\left(\frac{p\pi}{N}m\right) = \begin{cases} 0 \text{ if } p \neq n \\ N \text{ if } p = n \neq 0, N \\ 2N \text{ if } p = n = 0, N \end{cases} \tag{11.41}$$

Applying the orthogonality rules to the conditions on the partial derivatives specified in Equations 11.37 and 11.38 yields the following expressions for the coefficients.

$$a_0 = \frac{1}{2N}\sum_{m=0}^{2N-1} x_m \tag{11.42}$$

$$a_n = \frac{1}{N}\sum_{m=0}^{2N-1} x_m \cos\left(\frac{n\pi}{N}m\right) \, , 1 \leq n \leq N-1 \tag{11.43}$$

$$a_N = \frac{1}{2N}\sum_{m=0}^{2N-1} x_m \cos(\pi m) \tag{11.44}$$

$$b_n = \frac{1}{N}\sum_{m=0}^{2N-1} x_m \sin\left(\frac{n\pi}{N}m\right) \, , 1 \leq n \leq N-1 \tag{11.45}$$

The computational effort required to calculate the coefficients using the series given in Equations 11.42 to 11.45 is approximately proportional to N^2. Comparing this to the N^3 result given in Section 11.3 where a set of linear, algebraic equations was solved shows the inherent advantage to using the series solution, especially when N is large, as it usually is. For $N = 1000$, using the series method is approximately 1000 times faster than using the linear, algebraic equations.

11.5 Transforming to the Frequency Domain – The FFT

Up to this point, we have been attempting to maintain a geometric feeling for the transformations by describing the process as a curve fitting exercise using sines and cosines to

Table 11.1 Fourier Transforms Computational Effort.

Method	for $N = 2^P$	e.g. $p = 10$	Computer Time
Lin Alg Equations	2^{3p}	1.06×10^9	104,858
Series	2^{2p}	1.05×10^6	102
FFT	$p2^p$	1.02×10^4	1

represent the experimentally measured data. From a purely mathematical view, it is much better to do the analysis in the complex domain where the Fourier transform is given by

$$X(f, T) = \int_0^T x(t)e^{-i2\pi ft} dt \text{ where } i = \sqrt{-1} \tag{11.46}$$

Euler's relationship

$$e^{i\theta} = \cos\theta + i\sin\theta \tag{11.47}$$

shows how this is equivalent to the sine and cosine series we have been using.

The *Fast Fourier Transform* or *FFT* was developed in the mid-1960s. It is beyond the scope of this text to develop the actual algorithm. Suffice it to say that the computational effort is significantly less than other methods. The FFT algorithm requires that the number of data points be a power of two (i.e. $2N = 2^P$ in our cases where we were working with $2N$ points). The computational effort for the FFT is proportional to $N\log_2 N$, which is significantly less than for the series method presented in Section 11.4. For example, if $N = 2^{10} = 1024$, the computational effort of the three methods we have discussed is summarized in Table 11.1. The column labeled *Computer Time* shows the relative amount of computer processing time that the three methods would use on the same computer and is not in any particular unit of time. The table makes it abundantly clear why the FFT is the method of choice for all analysts. Implementations of the algorithm are commonplace and can even be found in spreadsheets.

11.6 Transforming to the Frequency Domain – An Example

As an example of the techniques for transforming to the frequency domain, consider the function

$$x(t) = 2 + 5\sin(2\pi t) + 10\sin(4\pi t) + 15\cos(6\pi t) \tag{11.48}$$

which is shown in Figure 11.3. This function has a mean value of 2 and three harmonics at frequencies of 1, 2, and 3 Hz. The harmonics have amplitudes of 5, 10, and 15 respectively.

Consider sampling the function at a rate of 10 samples per second ($\Delta t = 0.10$ s) and collecting 20 samples ($2N = 20$). The data are shown in Table 11.2. The base frequency for the transformation is given by Equation 11.27. That is,

$$f_0 = \frac{1}{2N\Delta t} = \frac{1}{20 \times 0.10} = 0.50 \text{ Hz} \tag{11.49}$$

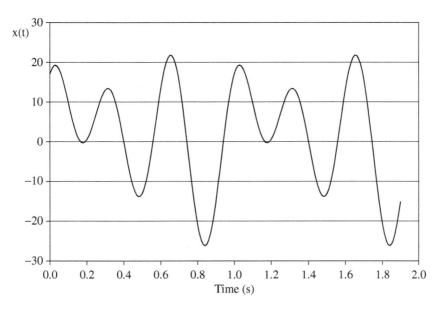

Figure 11.3 The example function, $x(t)$, plotted versus time.

Table 11.2 Sampled Data.

n	time (s)	x(t)
1	0.0	17.000
2	0.1	9.814
3	0.2	0.498
4	0.3	13.013
5	0.4	0.064
6	0.5	−13.000
7	0.6	13.207
8	0.7	15.258
9	0.8	−20.768
10	0.9	−15.085
11	1.0	17.000
12	1.1	9.814
13	1.2	0.498
14	1.3	13.013
15	1.4	0.064
16	1.5	−13.000
17	1.6	13.207
18	1.7	15.258
19	1.8	−20.768
20	1.9	−15.085

and we expect 20 coefficients, $a_0 \dots a_{10}$ and $b_1 \dots b_9$. Solving the set of linear, algebraic equations from Section 11.3 or using the series from Section 11.4 both result in the coefficients shown in Table 11.3. Notice that the FFT can't be used on this data set since the number of points is not a power of 2.

The original function can then be reconstructed using the coefficients in Table 11.3 and Equation 11.3. That is, retaining only the non-zero coefficients,

$$x_A(t) = a_0 + a_6 \cos(6\omega_0 t) + b_2 \sin(2\omega_0 t) + b_4 \sin(4\omega_0 t) \tag{11.50}$$

where

$$\omega_0 = 2\pi f_0 = 2\pi \times 0.50 = \pi \tag{11.51}$$

yielding

$$x_A(t) = 2.000 + 15.000 \cos(6\pi t) + 5.000 \sin(2\pi t) + 10.000 \sin(4\pi t) \tag{11.52}$$

which is exactly the same as the original function defined in Equation 11.48.

So it is clear that the transformation to the frequency domain can be undone and we can transform back to the time domain. This is called the *Inverse Fourier Transform*.

There is another way to write the DFT. We can use amplitudes and phase angles instead of the coefficients, a_n and b_n. We can write

$$a_n \cos(n\omega_0 t) + b_n \sin(n\omega_0 t) = A_n \cos(n\omega_0 t + \phi_n) \tag{11.53}$$

Table 11.3 DFT Coefficients.

Coefficient	Value	Frequency (Hz)
a_0	2.000	0.00
a_1	0.000	0.50
a_2	0.000	1.00
a_3	0.000	1.50
a_4	0.000	2.00
a_5	0.000	2.50
a_6	15.000	3.00
a_7	0.000	3.50
a_8	0.000	4.00
a_9	0.000	4.50
a_{10}	0.000	5.00
b_1	0.000	0.50
b_2	5.000	1.00
b_3	0.000	1.50
b_4	10.000	2.00
b_5	0.000	2.50
b_6	0.000	3.00
b_7	0.000	3.50
b_8	0.000	4.00
b_9	0.000	4.50

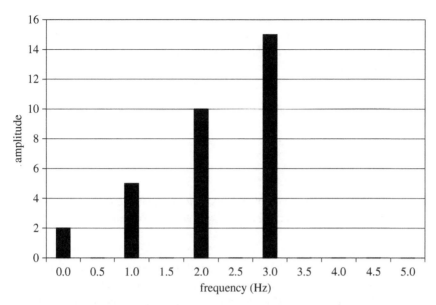

Figure 11.4 The DFT amplitudes of the example function, $x(t)$, plotted versus frequency.

where A_n is the amplitude

$$A_n = \sqrt{a_n^2 + b_n^2} \tag{11.54}$$

and ϕ_n is the phase angle

$$\phi_n = \arctan(b_n/a_n) \tag{11.55}$$

In fact, most data analysis concentrates on amplitudes only and phase angles are ignored. The function $x(t)$ that we have been considering (Equation 11.48 or 11.53) would appear on a plot of amplitude versus frequency as shown in Figure 11.4. We will be working strictly with amplitudes in the following.

11.7 Sampling and Aliasing

Consider the case where there is a single frequency harmonic signal at f Hz. That is

$$x(t) = A \cos(2\pi f t) + B \sin(2\pi f t) \tag{11.56}$$

If we used a DFT to detect this signal we would need to sample fast enough to capture the information required to calculate the coefficients a_n and b_n using Equations 11.43 and 11.45, for example. The question is *How fast must we sample to gather the necessary information about the signal?* The answer is fairly straightforward. You can visualize the process as follows.

Since there is only a single frequency present, all but one of the terms in the summation required for the DFT of Equation 11.29 will be zero and the DFT will reduce to

$$x(t) = a_p \cos(p\omega_0 t) + b_p \sin(p\omega_0 t) \tag{11.57}$$

where $p\omega_0 = 2\pi f$. That is, only the term at the frequency of the signal will remain.

Equation 11.57 has two unknowns, a_p and b_p. Returning to our curve fitting analogy, we have two unknowns so we will need two points per cycle in order to be able to calculate the amplitudes required. We therefore need to sample at a frequency at least twice as high as the frequency to be detected.

Let Δt be the time between samples. The sampling rate is then

$$f_s = \frac{1}{\Delta t} \text{Hz} \tag{11.58}$$

and the highest frequency that can be detected, called the *Nyquist frequency*, is

$$f_c = \frac{f_s}{2} = \frac{1}{2\Delta t} \text{Hz} \tag{11.59}$$

The obvious next question is *Since the Nyquist frequency is the highest that we can detect with our sampling rate, what happens if the signal has frequency content above the Nyquist frequency?*

Figure 11.5 shows a high-frequency signal (solid line) sampled at five points (squares) and the DFT approximation that would result (dashed line). Clearly, the high-frequency signal is in the data and the sampling will detect it but at less than two points per cycle so the DFT approximation will be a signal with a lower frequency. This is called *Aliasing*, a fitting name for a high-frequency signal masquerading as a low frequency signal.

In order to determine what frequency we can expect to see in the DFT, consider the following and note that the complex exponential notation for harmonics is used simply because it shortens the argument. Let there be a high-frequency signal, $x(t)$, with amplitude A, frequency ω, and phase angle ϕ so that we can write

$$x(t) = Ae^{i(\omega t + \phi)} = Ae^{i\omega t}e^{i\phi} \tag{11.60}$$

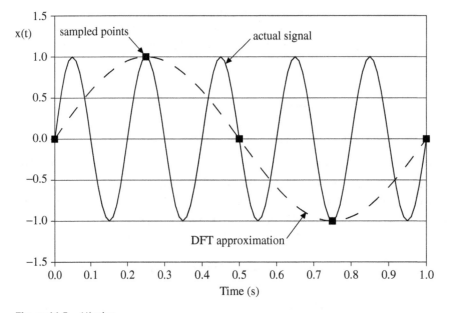

Figure 11.5 Aliasing.

and let the signal detected by the DFT, $x_d(t)$, have an amplitude a, be at a detected frequency, ω_d, and have a phase angle ϕ_d

$$x_d(t) = ae^{\iota(\omega_d t + \phi_d)t} = ae^{\iota\omega_d t}e^{\iota\phi_d} \tag{11.61}$$

where $\omega_d < \omega$.

Let there be a time t_0 where $x(t)$ and $x_d(t)$ are equal. This corresponds to one of the sampled points shown in Figure 11.5 and can be expressed as

$$Ae^{\iota\omega t_0}e^{\iota\phi} = ae^{\iota\omega_d t_0}e^{\iota\phi_d} \tag{11.62}$$

The two signals will be equal again at the next sampled point where $t = t_0 + \Delta t$ so we can write

$$Ae^{\iota\omega(t_0+\Delta t)}e^{\iota\phi} = ae^{\iota\omega_d(t_0+\Delta t)}e^{\iota\phi_d} \tag{11.63}$$

or

$$Ae^{\iota\omega t_0}e^{\iota\omega\Delta t}e^{\iota\phi} = ae^{\iota\omega_d t_0}e^{\iota\omega_d\Delta t}e^{\iota\phi_d} \tag{11.64}$$

Equation 11.62 can be used to cancel terms in Equation 11.64, resulting in

$$e^{\iota\omega\Delta t} = e^{\iota\omega_d\Delta t} \tag{11.65}$$

Define ω' to be the difference between ω and ω_d so that

$$\omega = \omega_d + \omega' \tag{11.66}$$

Substituting Equation 11.66 into Equation 11.65 yields

$$e^{\iota(\omega_d+\omega')\Delta t} = e^{\iota\omega_d\Delta t} \tag{11.67}$$

or

$$e^{\iota\omega_d\Delta t}e^{\iota\omega'\Delta t} = e^{\iota\omega_d\Delta t} \tag{11.68}$$

from which

$$e^{\iota\omega'\Delta t} = 1 \tag{11.69}$$

Equation 11.69 can be written as

$$\cos\omega'\Delta t + \iota\sin\omega'\Delta t = 1 \tag{11.70}$$

and, by equating the real and imaginary parts on the left and right-hand sides, we get two simultaneous equations, as follows

$$\cos\omega'\Delta t = 1 \quad\text{and}\quad \sin\omega'\Delta t = 0 \tag{11.71}$$

The dual requirements in Equation 11.71 are satisfied by

$$\omega'\Delta t = \pm 2k\pi; \, k = 0, 1, 2, \ldots \tag{11.72}$$

from which

$$\omega' = \pm 2k\pi\left(\frac{1}{\Delta t}\right); \, k = 0, 1, 2, \ldots \tag{11.73}$$

Substituting this into Equation 11.66 yields

$$\omega = \omega_d \pm 2k\pi \left(\frac{1}{\Delta t}\right); k = 0, 1, 2, \ldots \tag{11.74}$$

We can change Equation 11.74 from frequencies in rad/s to frequencies in Hz by noting that $\omega = 2\pi f$ and $\omega_d = 2\pi f_d$, resulting in

$$2\pi f = 2\pi f_d \pm 2k\pi \left(\frac{1}{\Delta t}\right); k = 0, 1, 2, \ldots \tag{11.75}$$

Canceling the 2π terms gives

$$f = f_d \pm k\left(\frac{1}{\Delta t}\right); k = 0, 1, 2, \ldots \tag{11.76}$$

We make one last change to Equation 11.76 by introducing the Nyquist frequency $f_c = 1/2\Delta t$ from Equation 11.59 to get

$$f = f_d \pm 2kf_c; k = 0, 1, 2, \ldots \tag{11.77}$$

Equation 11.77 specifies which high-frequency components, f, will appear at the detected frequency, f_d. It is clear that, since $f_d \le f_c$, some of the high frequencies will be negative. These frequencies are actually positive but correspond to a phase change in the detected signal. The positive frequencies are

$$f = f_d + 2kf_c; k = 0, 1, 2, \ldots \tag{11.78}$$

and

$$f = -f_d + 2kf_c; k = 0, 1, 2, \ldots \tag{11.79}$$

Equations 11.78 and 11.79 can be combined as

$$f = 2kf_c \pm f_d; k = 0, 1, 2, \ldots \tag{11.80}$$

The Nyquist frequency is sometimes called the *folding frequency* because of Equation 11.80. For the case where $k = 1$, we find that the first frequency aliased with f_d is

$$f = 2f_c - f_d \tag{11.81}$$

and this frequency can be found by folding the frequency scale shown in Figure 11.6 around the Nyquist frequency, f_c.

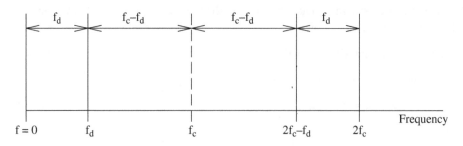

Figure 11.6 The folding frequency.

Consider again the example of Section 11.6 but, in addition to the signal, $x(t)$, that has three frequency components, we will add another higher-frequency component at 8.5 Hz. The sampling time is $\Delta t = 0.10$ s, so that the highest frequency that will be detectable is the Nyquist Frequency $f_c = 1/(2\Delta t) = 5$ Hz. As a result, the 8.5 Hz signal is outside the detectable range but we expect to see it as an aliased signal at (according to Equation 11.81) $f_d = 2 \times 5 - 8.5 = 1.5$ Hz. Let the 8.5 Hz component have an amplitude of 20 so that the function in Equation 11.48 becomes

$$x(t) = 2 + 5\sin(2\pi t) + 10\sin(4\pi t) + 15\cos(6\pi t) + 20\cos(17\pi t) \qquad (11.82)$$

Table 11.4 shows the coefficients that are calculated by the DFT for this revised case. Notice that a_3, the coefficient related to the cosine term at 1.5 Hz, erroneously has the amplitude of the 8.5 Hz component. Aliasing is a serious problem in experimental data analysis. Figure 11.7 shows the dramatic change from Figure 11.4.

There will invariably be signal components at frequencies higher than the frequency range of interest in the experiment. The only way to avoid aliasing is to use a low-pass analog filter when the data are recorded. The filtering has to be done using hardware. Section 11.9 discusses digital filtering after the data have been recorded but it needs to be emphasized that digital filtering relies on accurate DFTs and, if the aliased results are in the data,

Table 11.4 DFT Coefficients.

Coefficient	Value	Frequency (Hz)
a_0	2.000	0.00
a_1	0.000	0.50
a_2	0.000	1.00
a_3	**20.000**	1.50
a_4	0.000	2.00
a_5	0.000	2.50
a_6	15.000	3.00
a_7	0.000	3.50
a_8	0.000	4.00
a_9	0.000	4.50
a_{10}	0.000	5.00
b_1	0.000	0.50
b_2	5.000	1.00
b_3	0.000	1.50
b_4	10.000	2.00
b_5	0.000	2.50
b_6	0.000	3.00
b_7	0.000	3.50
b_8	0.000	4.00
b_9	0.000	4.50

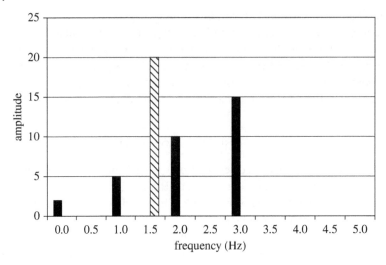

Figure 11.7 Aliased DFT results.

the digital filtering process will not remove them because they are seen to be legitimate low frequency signals.

11.8   Leakage and Windowing

To illustrate the concept of *leakage*, we consider an example where we have a signal with two frequencies. Let

$$x(t) = 60\cos(40\pi t) + 90\cos(80\pi t) \tag{11.83}$$

That is, we have an amplitude of 60 at 20 Hz and an amplitude of 90 at 40 Hz. Choose $\Delta t = 0.01$s and collect 100 samples. As a result we have a sample time $T = 100 \times 0.01 = 1$ second, a frequency resolution $\Delta f = 1/T = 1$ Hz, and a Nyquist frequency $f_c = 1/(2\Delta t) = 50$ Hz. The DFT is calculated and the results, shown in Figure 11.8, are as expected.

Now we consider a different signal with the same amplitudes but slightly different frequencies (amplitude of 60 at 20.5 Hz and an amplitude of 90 at 39.5 Hz)

$$x(t) = 60\cos(41\pi t) + 90\cos(79\pi t) \tag{11.84}$$

Using the same settings for the DFT as in the first signal, the result is as shown in Figure 11.9. The immediate question is *What happened?*

Before answering the *What happened?* question, it must be pointed out that we are looking at what is called *Leakage*. This is a phenomenon where the amplitudes that are present in the signal at discrete frequency bins *leak* into neighboring bins.

As to the question of *What happened?*, we can start by considering the continuous Fourier transform of a square wave. Figure 11.10 shows the square wave and some of the components that are used in approximating it. The thin lines are components in the expansion and the thick line is the sum of the components. Theoretically, an infinite number of components will exactly duplicate the square wave but, as can already be seen, this will require

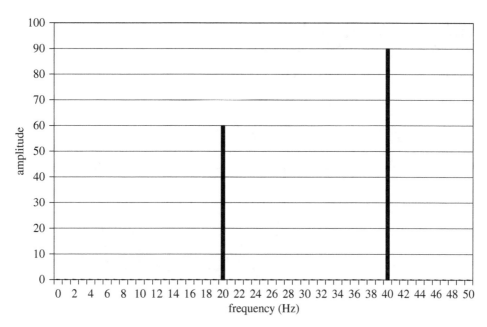

Figure 11.8 The DFT for the first example.

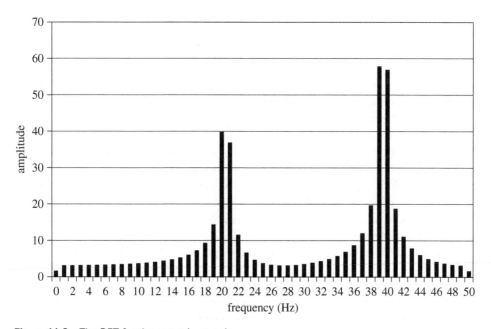

Figure 11.9 The DFT for the second example.

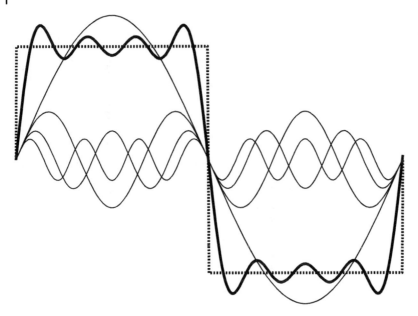

Figure 11.10 CFT approximation to the square wave.

some amplitude at every frequency in order to fill in the discontinuities. The square wave has the same period as the first component of the Fourier series but the corners can only be filled in by adding higher and higher-frequency components.

Going back to our two examples, the difference is that the first example (20 Hz and 40 Hz) is a signal where both components have completed an integer number of cycles in the 1 second sample that we considered. In the second example (20.5 Hz and 39.5 Hz), neither component is at the end of a cycle when the sample is truncated so the DFT sees a discontinuity and tries to deal with it by using all available frequency components.

The Fourier transform is only valid for a periodic function and requires information for complete cycles of the signal. The chances of any experimental data sample being periodic and being sampled over complete cycles is essentially zero. Leakage can't be completely eliminated but it can be reduced by *Windowing* the data. Windowing forces the data to appear to be periodic by multiplying them by a function that is zero at both the start and the end and that rises to a value of one in the center.

You can imagine all sorts of windows. For example, the Rectangular window goes from zero to one at the beginning of the data, stays at one, and then goes back down to zero at the end of the data. This is the default window that we get without manipulating the data so it has no effect on leakage. The simplest window that has an effect is the Triangular window that ramps from zero at the start of the data to one at the center of the data and then back down to zero again at the end of the data. It is, however, more common to use windows that have continuous derivatives and zero slopes at their ends. A very common window function is the Hanning window shown in Figure 11.11

The Hanning window is expressed as a weighting function, $w(x)$

$$w(x) = \frac{1}{2}(1 + \cos \pi x); \quad -1 \leq x \leq 1 \tag{11.85}$$

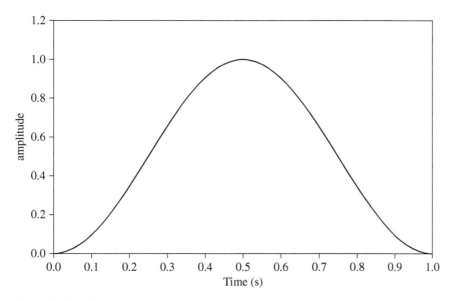

Figure 11.11 The Hanning window.

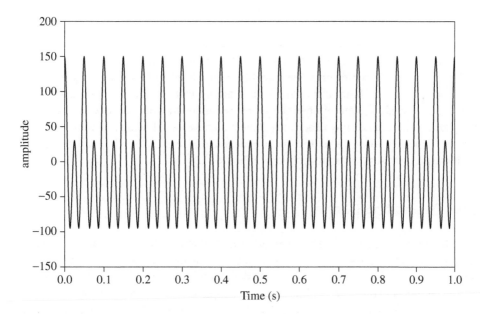

Figure 11.12 The data from Equation 11.83.

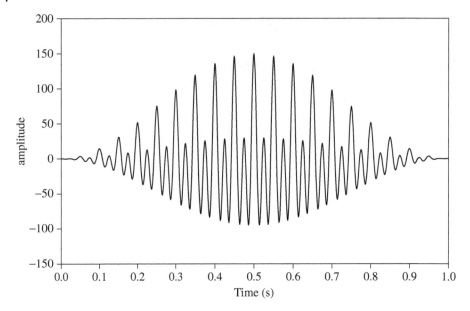

Figure 11.13 The windowed data.

and is shown time-scaled in Figure 11.11 in order to cover the one second of data we have been considering in our examples.

Figures 11.12 and 11.13 show the original data and the windowed data respectively. It is clear from the change in shape of the data that the windowing will have an effect on the amplitudes of the DFT coefficients. We take the amplitude effect into account by realizing that the area under the Hanning window is one-half the area under a rectangular window. As a result, the windowed amplitudes need to be multiplied by two in order to restore their actual values. This area scaling needs to be done whichever windowing function is used.

Figure 11.14 shows the DFT for the windowed data of Figure 11.13. While not being perfect, the result is far superior to that shown in Figure 11.9.

11.9 Decimating Data

There are times when we realize that we have sampled at a higher rate than is necessary and we wish to reduce the sampling rate. For example (and this has happened), if we set the sampling rate in our A/D converter to 100,000 samples per second and we analyze using an FFT limited to 8192 points, we will have a total sample time $T = 0.08192$ seconds and the frequency resolution will then be $\Delta f = 1/T = 12.2$ Hz. If your analysis is meant to be able to distinguish between signals at 50 and 60 Hz, for instance, then you are out of luck because both of these signals will be in the same frequency bin. It becomes tempting to say that, since this is all digital data, we can simply ignore some of the data and we will get the same result as if we had sampled more slowly. That is, just using every second point will be equivalent to sampling at 50,000 samples per second or, even better, using every hundredth point will be equivalent to sampling at 1000 Hz so that 8192 points will give $\Delta f = 0.122$ Hz.

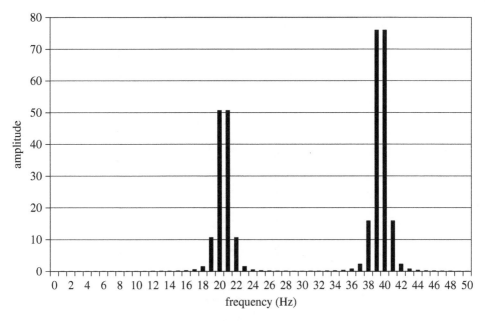

Figure 11.14   The DFT for the second example with windowing.

While it is true that we can leave out data points and the result will be exactly as if we had sampled at a slower rate, we need to keep in mind the possibility of aliasing as discussed in Section 11.7. If we did our work correctly when we sampled the data, we used a low-pass analog filter with a cut-off frequency we desired and then sampled at a rate that was at least twice as high as the cut-off frequency. For the example just discussed, we sampled at 100,000 samples per second so our cut-off frequency would have been around 50,000 Hz. The digital signal we recorded therefore has frequency components up to 50,000Hz. If we then decide to use every hundredth point to set our sampling rate to 1000 Hz, we are going to get serious aliasing.

The process of leaving out data points is called *downsampling*.

This aliasing situation is illustrated in Figures 11.15 and 11.16. Figure 11.15 shows the same example we used for leakage (Equation 11.83 and Figure 11.8) where we had signals at 20 and 40 Hz and we used 100 samples with a time between samples $\Delta t = 0.01$ seconds, giving a Nyquist frequency of 50 Hz. Figure 11.16 shows the results of using only every second data point so that $\Delta t = 0.02$ seconds and the Nyquist frequency is 25 Hz. The peak that was at 40 Hz is now above the Nyquist frequency and is aliased to appear at 10 Hz.

There is a difference between the aliasing stemming from downsampling and that which comes from not using a low-pass filter when sampling data. If you omit the low-pass filter your DFT will interpret high-frequency signals as low frequency signals and you won't be able to differentiate between actual peaks and aliased peaks. In the case of downsampling, you know where the aliased peaks actually are in the sense that you see them in the DFT before you do the downsampling. This gives you the option of using a digital filter to remove them before you do the downsampling. The sequence of using a low-pass digital filter followed by downsampling is called *decimation*.

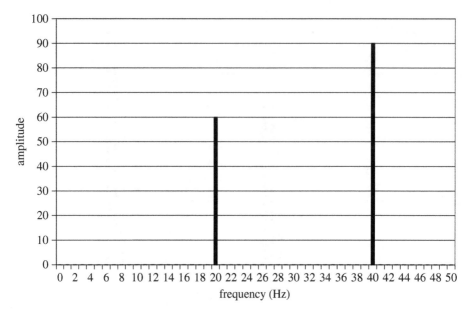

Figure 11.15 The DFT before downsampling.

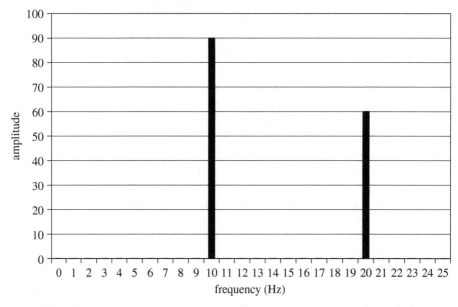

Figure 11.16 The DFT after downsampling.

For the example we are using, we want to get rid of anything above 25 Hz before we down-sample. This is best accomplished in the frequency domain by performing a DFT on the full data set, then implementing a low-pass filter with a cut-off at 25 Hz, and then performing the inverse DFT to get a revised time signal. Implementing the filter in the frequency domain is simply a matter of using a rectangular window that is equal to one for frequencies below the cut-off frequency and zero above it. This can be accomplished by setting the DFT coefficients related to frequencies above the cut-off frequency to zero.

Figure 11.17 shows the time series before (solid line) and after (dashed line) digital filtering with a cut-off frequency of 25 Hz. It is clear that the higher-frequency component of the signal (the 40 Hz component) has been completely eliminated by the filter and only the 20 Hz component remains. Figure 11.18 shows the result of sampling the new signal at $\Delta t = 0.02$ seconds and performing the DFT.

There is also the possibility of performing digital filtering in the time domain. The techniques used are found in works related to *Digital Signal Processing* or DSP and are basically simulations of hardware filters. The results are not as good as those of the frequency domain filters just discussed but the frequency domain filters are sometimes not viable when there are very large data sets or data is streaming in real time.

Consider the RC circuit shown in Figure 11.19. This is a traditional low-pass filter circuit. It is not the best hardware filter to use but is useful here as an example. Analysis of this circuit is based on the currents flowing through the resistor and the capacitor being equal. The current passing through the resistor is

$$i = \frac{1}{R}(V_{in} - V_{out}) \tag{11.86}$$

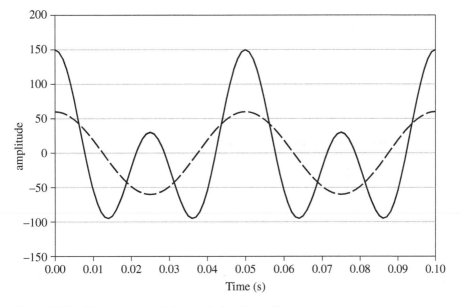

Figure 11.17 The time series before and after digital filtering.

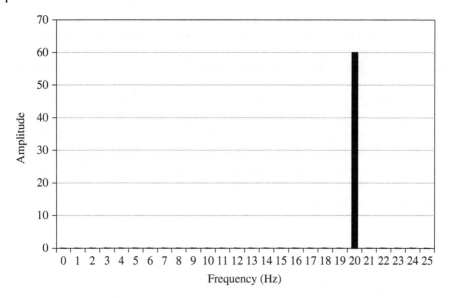

Figure 11.18 The DFT after decimating.

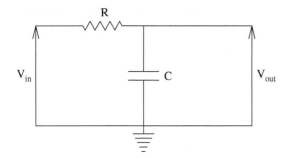

Figure 11.19 A low-pass filter circuit.

and through the capacitor is

$$i = C\frac{dV_{out}}{dt}$$

(11.87)

Equating the currents gives the differential equation

$$C\frac{dV_{out}}{dt} = \frac{1}{R}(V_{in} - V_{out})$$

(11.88)

which can be rewritten as

$$RC\frac{dV_{out}}{dt} + V_{out} = V_{in}$$

(11.89)

Let V_{in} be a harmonic function with frequency ω. That is

$$V_{in} = V_i e^{i\omega t}$$

(11.90)

Then, since Equation 11.89 is a linear, first-order, ordinary, differential equation, the solution will be

$$V_{out} = V_o e^{\iota \omega t} \tag{11.91}$$

so that

$$\frac{dV_{out}}{dt} = \iota \omega V_o e^{\iota \omega t} \tag{11.92}$$

Substituting Equations 11.90 through 11.92 into Equation 11.89 gives

$$(\iota RC\omega + 1)V_o e^{\iota \omega t} = V_i e^{\iota \omega t} \tag{11.93}$$

from which the amplitude ratio can be found to be

$$\frac{V_o}{V_i} = \frac{1}{(\iota RC\omega + 1)} \tag{11.94}$$

The amplitude ratio as written in Equation 11.94 is complex and therefore contains both amplitude and phase information. We are interested in only the amplitude information so we write

$$\left| \frac{V_o}{V_i} \right| = \frac{1}{\sqrt{R^2 C^2 \omega^2 + 1}} \tag{11.95}$$

We define the cut-off frequency (ω_c) to be the point at which the filter attenuates the signal to one-half its unfiltered power. That is, since power is proportional to voltage squared, we let

$$\left(\frac{V_o}{V_i} \right)^2 = \frac{1}{R^2 C^2 \omega_c^2 + 1} = \frac{1}{2} \tag{11.96}$$

yielding

$$\omega_c = \frac{1}{RC} \tag{11.97}$$

and we can rewrite Equation 11.95 as

$$\left| \frac{V_o}{V_i} \right| = \frac{1}{\sqrt{\left(\frac{\omega}{\omega_c} \right)^2 + 1}} \tag{11.98}$$

The signal attenuation provided by the RC filter (Equation 11.98) is most often presented in the literature on a plot where the attenuation is in Decibels (dB $= 20 \log_{10}$) and the frequency ratio, (ω/ω_c), is on a logarithmic scale. Figure 11.20 shows this plot for the low-pass RC filter. Figure 11.21 shows the attenuation factor on a linear scale. In both figures, the dashed vertical line is the cut-off frequency. The linear scale gives a better feeling for the actual attenuation. The signal is attenuated over the entire frequency range and only 70% of the signal remains at the cut-off frequency.

We now turn to the problem of implementing a digital filter in the time domain. We start with the differential equation governing the RC low-pass filter (Equation 11.89) and use an

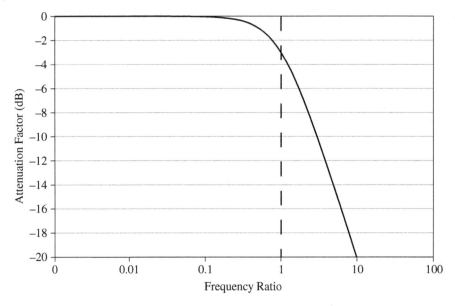

Figure 11.20 Low-pass filter frequency response in dB on a logarithmic scale.

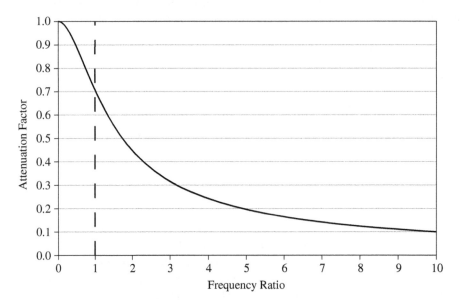

Figure 11.21 Low-pass filter frequency response on a linear scale.

Euler approximation to the first derivative using a backwards difference over the sampling time Δt

$$\frac{dV\,(t)_{out}}{dt} \approx \frac{V\,(t)_{out} - V\,(t - \Delta t)_{out}}{\Delta t} \tag{11.99}$$

Substitute this into Equation 11.89 to get

$$RC \left[\frac{V\ (t)_{out} - V\ (t - \Delta t)_{out}}{\Delta t} \right] + V\ (t)_{out} = V\ (t)_{in} \tag{11.100}$$

This can be rearranged to give the filtered output at time t based on the input at time t and the filtered output at time $t - \Delta t$ as

$$V\ (t)_{out} = \left[\frac{\Delta t}{RC + \Delta t} \right] V\ (t)_{in} + \left[\frac{RC}{RC + \Delta t} \right] V\ (t - \Delta t)_{out} \tag{11.101}$$

Equation 11.101 can be written in a more general form if we define a parameter α as

$$\alpha = \left[\frac{\Delta t}{RC + \Delta t} \right] \tag{11.102}$$

which results in

$$\left[\frac{RC}{RC + \Delta t} \right] = (1 - \alpha) \tag{11.103}$$

so that Equation 11.101 can be written as

$$V\ (t)_{out} = \alpha V\ (t)_{in} + (1 - \alpha)V\ (t - \Delta t)_{out} \tag{11.104}$$

Equation 11.104 is the expression for the *Exponential Moving Average* that is widely used to smooth time series data in fields such as Economics. The parameter α is called a *smoothing factor*.

Applying this smoothing technique to the example we have been using (amplitudes 60 at 20 Hz and 90 at 40 Hz, cut-off frequency of 25 Hz, $\Delta t = 0.02$ s) results in the smoothed data plot shown as Figure 11.22, where the solid line is the original signal and the dashed line is

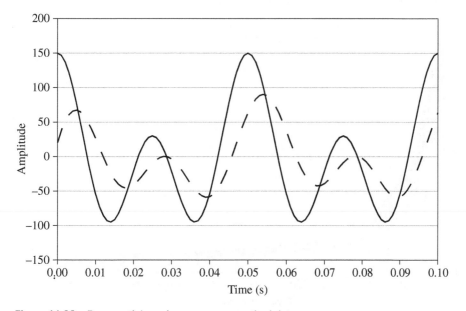

Figure 11.22 Exponential moving average smoothed data.

the smoothed data. The DFT plot resulting from the smoothed data is shown as Figure 11.23. Clearly the effect is not nearly as good as that obtained from the frequency domain filter.

The frequency response of the digital filter is shown in Figures 11.24 and 11.25. Also shown in the figures is a heavy dark line that is the theoretical analog frequency response for the same filter as calculated from Equation 11.98. The approximation is relatively good but the numerical problems demonstrated by the waviness are apparent.

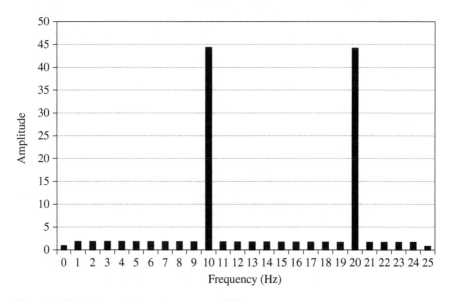

Figure 11.23 Exponential moving average DFT.

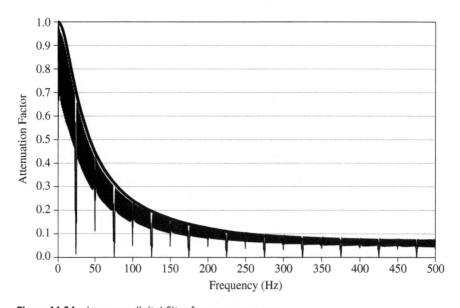

Figure 11.24 Low-pass digital filter frequency response.

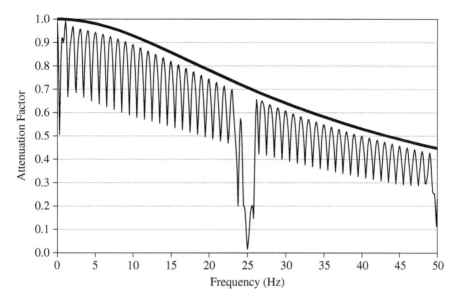

Figure 11.25 Low-pass filter frequency response near the cut-off frequency.

11.10   Averaging FFTs

The last topic to be covered in this chapter has to do with suppression of noise in experimental measurements. Noise in measurements consists of random changes in signal values that can, in some cases, reach levels that approach that of the signal you are trying to measure. It is good experimental practice to attempt to minimize noise levels before recording data. Nevertheless, there will always be some level of noise in experimental data.

Figure 11.26 shows a noisy time signal. The signal is, in fact, the same signal that we have been using as an example (amplitudes 60 at 20 Hz and 90 at 40 Hz) with pseudo-random noise distributed between amplitudes of −150 and +150 superimposed on it. Figure 11.27 shows the original smooth data as a dashed line and the resulting noisy data as a solid line for a small interval of time.

Since the signal comprises two components, one of which is regular and repeatable while the other is random, it can be expected that two DFTs calculated over different parts of the time signal will give different results. Both will have large amplitudes at the frequencies of the signal but smaller and non-repeatable amplitudes at other frequencies. It can therefore be expected that if several of the DFTs are averaged, the real signal will be prominent since it appears in every DFT whereas the noise will give different peaks in different spectra and, on average, will not be prominent. This is exactly what happens. The noise is part of the signal and will never be zero, but averaging the spectra results in a smoothing of the noise. This is a simple linear averaging process where the DFT amplitudes at each frequency are added and then divided by the number of DFTs performed.

Figures 11.28, 11.29, and 11.30 show the DFTs for 1, 10, and 20 averages respectively. The smoothing of the noise is evident.

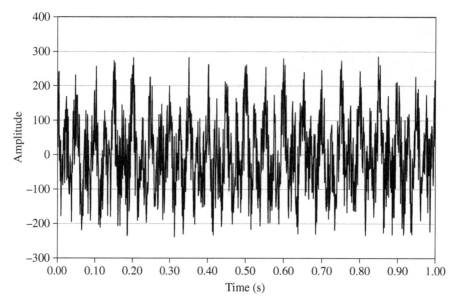

Figure 11.26 Noisy time signal.

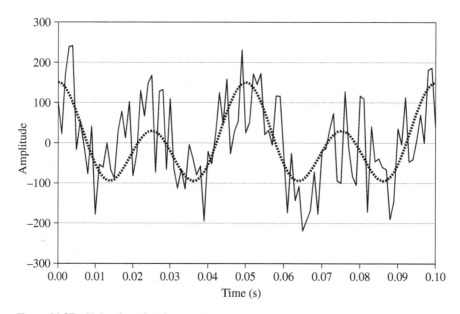

Figure 11.27 Noisy time signal zoomed.

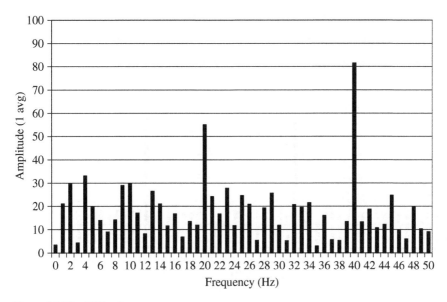

Figure 11.28   DFT – 1 average.

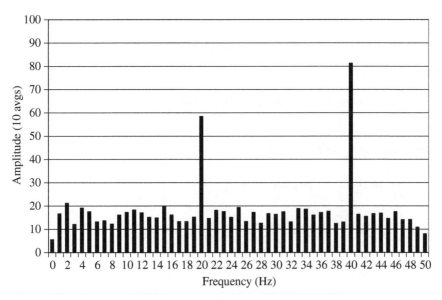

Figure 11.29   DFT – 10 averages.

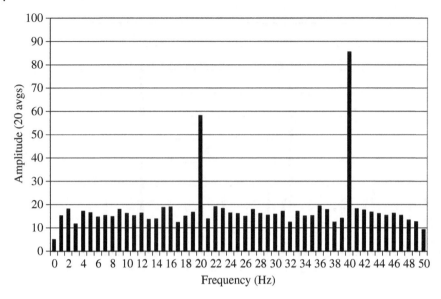

Figure 11.30 DFT – 20 averages.

Exercises

11.1 Prove that the trigonometric integrals in Equation 11.10 are zero. That is, for $\omega_0 = 2\pi/T$, show that:

$$\int_0^T \cos(n\omega_0 t)dt = 0 \text{ and } \int_0^T \sin(n\omega_0 t)dt = 0$$

11.2 Show that, for $\omega_0 = 2\pi/T$ and $p \neq n$,

$$\int_0^T \cos(n\omega_0 t)\sin(p\omega_0 t)dt = 0$$

$$\int_0^T \sin(n\omega_0 t)\sin(p\omega_0 t)dt = 0$$

$$\int_0^T \cos(n\omega_0 t)\cos(p\omega_0 t)dt = 0$$

$$\int_0^T \sin(n\omega_0 t)\cos(p\omega_0 t)dt = 0$$

by using the identities

$$\cos(n\omega_0 t)\cos(p\omega_0 t) = \frac{1}{2}\cos[(n+p)\omega_0 t] + \frac{1}{2}\cos[(n-p)\omega_0 t]$$

$$\sin(n\omega_0 t)\sin(p\omega_0 t) = \frac{1}{2}\cos[(n-p)\omega_0 t] - \frac{1}{2}\cos[(n+p)\omega_0 t]$$

$$\cos(n\omega_0 t)\sin(p\omega_0 t) = \frac{1}{2}\sin[(n+p)\omega_0 t] - \frac{1}{2}\sin[(n-p)\omega_0 t]$$

$$\sin(n\omega_0 t)\cos(p\omega_0 t) = \frac{1}{2}\sin[(n+p)\omega_0 t] + \frac{1}{2}\sin[(n-p)\omega_0 t]$$

11.3 Use Equation 11.85 to show that the area under the Hanning Window is one-half the area under the Rectangular Window.

11.4 Your data acquisition system can sample at any rate you choose between 1 and 4000 samples per second and it has sufficient memory to store up to 8192 samples. You also have a low-pass filter with an adjustable cut-off frequency that can be set anywhere between 1 and 2000 Hz. You have been asked to come into a building experiencing excessive floor vibrations and determine whether the cause is a rotary machine operating at 1800 RPM in this building or a reciprocating compressor operating at 28 strokes per second in an adjacent building.

You have an accelerometer and plan to measure and record accelerations on the floor. You will then perform an FFT and calculate the response spectrum to see which machine is at fault.

Specify your sampling rate, how many samples you should take per measurement, what your cut-off frequency should be, and how many measurements you can average. Give reasons for your choices.

11.5 A company is having a vibration problem. They measured the acceleration versus time plot that is shown below and have come to you for advice. They tell you that they used a sampling rate of 100 samples/second and they had a low-pass filter with a cut-off frequency of 50 Hz.

(a) Explain the nature of the problem that the company is having.
(b) You want to run an FFT on the data to see what frequencies are present. Working with the plot below, estimate the frequency resolution, in Hz, that you would need.
(c) How many points will you use for the FFT and what frequency resolution will result? How many averages are possible? What will be the maximum frequency you can get?

11.6 Program the DFT series of Section 11.4 using software of your choice and then generate some data similar to that used as an example in Section 11.6 (i.e. Equation 11.48) and test your software on it. Experiment with the software – try out aliasing, leakage, and so on.

11.7 If you install strain gauges on the spindle of a milling machine, you will be able to pick up both the relatively constant normal force on the cutter and the cyclic forces caused by tooth engagement as the spindle turns. If you plot the total force versus time, you will see the cyclic forces as a relatively small oscillation superimposed on the large, constant, normal force. The amplitude of the cyclic force variation is about 5% of that of the normal force.

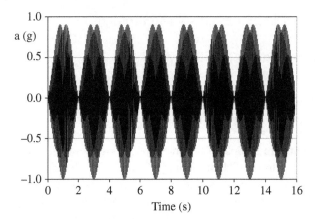

Figure E11.5

(a) What is the mean value of this signal?
(b) What would you expect the DFT of the signal to look like?
(c) If you calculated the mean value in the time domain and subtracted it from each of the data points before doing the DFT, what would your DFT look like?
(d) Using the DFT of the zero mean signal in part (c) is the preferred way of getting the frequency content of a signal in most cases. Why do you think this is?

12

Topics in Vibrations

Here we will consider various topics in vibrations that require the basic knowledge gained from the previous material. Essentially, we will be discussing applications of vibrations to various problems.

12.1 What About the Mass of the Spring?

We start with a question that has likely crossed the mind of everyone who has ever studied vibrations. That is, we are continually modeling systems using lumped mass and lumped stiffness, as shown in Figure 12.1, when we all know that springs can be very heavy. How do we take the mass of the spring into account in the model?

One way to address this is by finding the kinetic energy contributed by the spring and adding that to the kinetic energy contribution of the mass before using Lagrange's Equation. We do this as follows.

Let the spring have a total mass, m_s, and a total length, ℓ. Now consider that the displacement of a point on the spring is zero at the ground and is equal to the motion of the mass, x, where the spring is attached to the mass. Since we are working with a linear spring, we assume that the displacement varies linearly along its length so we can write the local displacement, $\delta(y)$ as

$$\delta(y) = \frac{y}{\ell} x \tag{12.1}$$

where y measures the distance from the ground to the point on the spring ($0 \leq y \leq \ell$).

Let there be a small element of the spring at position y that has length, dy. The mass of this element will be

$$dm = \frac{m_s}{\ell} dy \tag{12.2}$$

We can differentiate Equation 12.1 to get the velocity of the element as

$$\dot{\delta}(y) = \frac{y}{\ell} \dot{x} \tag{12.3}$$

Introduction to Mechanical Vibrations, First Edition. Ronald J. Anderson.
© 2020 John Wiley & Sons Ltd. Published 2020 by John Wiley & Sons Ltd.
Companion website: www.wiley.com/go/anderson/introduction-to-vibrations

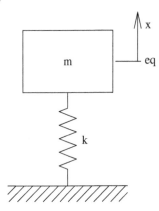

Figure 12.1 A mass on a spring.

and then the contribution of the element to the kinetic energy is

$$dT_s = \frac{1}{2} dm \, \dot{\delta}^2(y) = \frac{1}{2} \left(\frac{m_s}{\ell} dy\right) \left(\frac{y}{\ell}\dot{x}\right)^2 = \frac{1}{2} \left(\frac{m_s \dot{x}^2}{\ell^3}\right) y^2 dy \qquad (12.4)$$

Integrating over the length of the spring gives the total contribution of the spring to the kinetic energy as

$$T_s = \frac{1}{2} \left(\frac{m_s \dot{x}^2}{\ell^3}\right) \int_0^\ell y^2 dy = \frac{1}{2} \left(\frac{m_s \dot{x}^2}{\ell^3}\right) \left[\frac{y^3}{3}\right]_0^\ell = \frac{1}{2} \left(\frac{m_s \dot{x}^2}{\ell^3}\right) \left(\frac{\ell^3}{3}\right) \qquad (12.5)$$

which simplifies to

$$T_s = \frac{1}{2} \left(\frac{m_s}{3}\right) \dot{x}^2 \qquad (12.6)$$

Adding this to the kinetic energy contribution of the mass, we get a total kinetic energy of

$$T = \frac{1}{2} m\dot{x}^2 + \frac{1}{2} \left(\frac{m_s}{3}\right) \dot{x}^2 = \frac{1}{2} \left(m + \frac{m_s}{3}\right) \dot{x}^2 \qquad (12.7)$$

so the answer to the question of what to do with the mass of the spring is that we add one third of the spring mass to the lumped mass.

This analysis gives a "rule of thumb" for dealing with the spring mass but is only correct for the case where one end of the spring is connected to the ground. In the case where the spring is attached to two moving bodies, as in Figure 12.2, a similar analysis goes as follows.

First, Equation 12.1 for the local displacement of a point on the spring needs to be modified to include x_1, giving

$$\delta(y) = x_1 + \frac{y}{\ell}(x_2 - x_1) = x_1 \left(1 - \frac{y}{\ell}\right) + x_2 \left(\frac{y}{\ell}\right) \qquad (12.8)$$

where y now measures the distance from the spring connection on m_1 to the point on the spring ($0 \le y \le \ell$).

The velocity of the element of the spring located at y is

$$\dot{\delta}(y) = \dot{x}_1 \left(1 - \frac{y}{\ell}\right) + \dot{x}_2 \left(\frac{y}{\ell}\right) \qquad (12.9)$$

and the contribution of the element to the total kinetic energy becomes

$$dT_s = \frac{1}{2} \left(\frac{m_s}{\ell}\right) \left[\dot{x}_1^2 \left(1 - \frac{y}{\ell}\right)^2 + 2\dot{x}_1\dot{x}_2 \left(1 - \frac{y}{\ell}\right) \left(\frac{y}{\ell}\right) + \dot{x}_2^2 \left(\frac{y}{\ell}\right)^2\right] dy \qquad (12.10)$$

Figure 12.2   A spring connecting two masses.

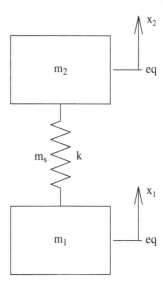

Integrating from 0 to ℓ gives, after some simplification, the total contribution of the spring to the kinetic energy as

$$T_s = \frac{1}{2}\left(\frac{m_s}{3}\right)\dot{x}_1^2 + \left(\frac{m_s}{6}\right)\dot{x}_1\dot{x}_2 + \frac{1}{2}\left(\frac{m_s}{3}\right)\dot{x}_2^2 \qquad (12.11)$$

The kinetic energy of the two masses can be added to this to give the total kinetic energy

$$T = \frac{1}{2}\left(m_1 + \frac{m_s}{3}\right)\dot{x}_1^2 + \left(\frac{m_s}{6}\right)\dot{x}_1\dot{x}_2 + \frac{1}{2}\left(m_2 + \frac{m_s}{3}\right)\dot{x}_2^2 \qquad (12.12)$$

The inertial terms in the equations of motion will then be

$$\frac{d}{dt}\left(\frac{\partial T}{\partial \dot{x}_1}\right) = \left(m_1 + \frac{m_s}{3}\right)\ddot{x}_1 + \left(\frac{m_s}{6}\right)\ddot{x}_2 \qquad (12.13)$$

and

$$\frac{d}{dt}\left(\frac{\partial T}{\partial \dot{x}_2}\right) = \left(\frac{m_s}{6}\right)\ddot{x}_1 + \left(m_2 + \frac{m_s}{3}\right)\ddot{x}_2 \qquad (12.14)$$

where we can see that the "rule of thumb" is still approximately correct in that we add one third of the mass of the spring to each body but we also need to look after the inertial coupling between x_1 and x_2 that is indicated by the one sixth of the spring mass appearing in the equations[1].

12.2   Flow-induced Vibrations

We consider now a case of *self-excited vibrations*. This is one example of a forced vibration problem where the external forces are not harmonic but in which the amplitudes of motion

---

1   Equation 9.65 gives the mass matrix for the rod element, which is exactly what we have just derived here. ρAL is the total mass of the rod element.

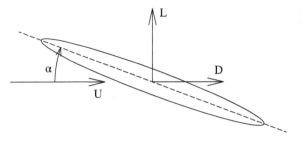

Figure 12.3 Lift and drag in a wind tunnel.

can grow without bound because the external forces are continually adding energy to the system[2].

Figure 12.3 shows an airfoil in a steady air flow of velocity U inside a wind tunnel. Wind tunnel tests are undertaken to measure the lift force, L, and the drag force, D, as a function of the angle of attack, α. This is done by rotating the airfoil to a specified angle of attack and then using force transducers to measure L and D. Some typical results for an elliptical airfoil are shown in Figure 12.4.

Figure 12.5 shows the same airfoil, now having a mass, m, and supported by the spring of stiffness, k. We let the body have a single DOF, x, measured positively upwards as shown in the figure. The air flows horizontally at speed, U, and the airfoil has vertical velocity, $\dot{x}$, upwards so that the velocity of the air relative to the body has a downward component and an angle of attack is generated as shown in the figure where V_{rel} shows the velocity of the air relative to the body. Comparing the directions in Figure 12.5 to those of the wind tunnel

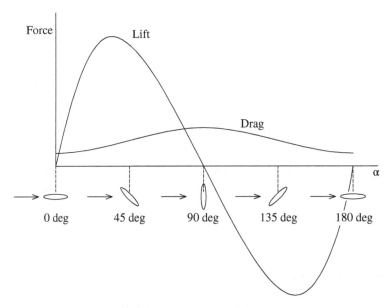

Figure 12.4 Typical lift and drag forces versus angle of attack.

---

2 This analysis was first published by J.P. Den Hartog, Professor of Mechanical Engineering at the Massachusetts Institute of Technology, in 1932. It can be found in the book *Mechanical Vibrations*, by J.P Den Hartog, 4th Edition, McGraw-Hill Book Company, 1956

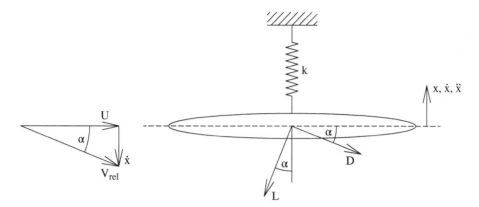

Figure 12.5   A suspended airfoil in steady flow.

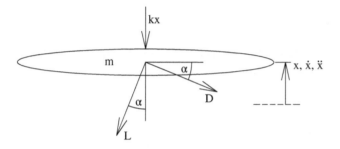

Figure 12.6   Free body diagram of the airfoil.

test in Figure 12.3, we can see that the directions of L and D must be as shown. We can write the angle of attack as

$$\alpha = tan^{-1} \frac{\dot{x}}{U} \tag{12.15}$$

A vertical force balance on the mass as shown in the free body diagram of Figure 12.6 yields

$$-L \cos \alpha - D \sin \alpha - kx = m\ddot{x} \tag{12.16}$$

We can define $F(\alpha) = L \cos \alpha + D \sin \alpha$ and rearrange Equation 12.16 to get

$$m\ddot{x} + kx = -F(\alpha) \tag{12.17}$$

There will be an equilibrium position of the body where $x = x_0$ and $\dot{x} = \ddot{x} = 0$. Also note that Equation 12.15 gives $\alpha = 0$ when $\dot{x} = 0$. The equilibrium condition is then

$$kx_0 = -F(0) \tag{12.18}$$

In the usual fashion, we consider small motions away from equilibrium by defining a new displacement variable, y and writing $x = x_0 + y$, $\dot{x} = \dot{y}$, and $\ddot{x} = \ddot{y}$. Note that the angle of attack in Equation 12.15 can be approximated as

$$\alpha \approx \frac{\dot{y}}{U} \tag{12.19}$$

since the tangent of a small angle is approximately equal to the angle and $\dot{y}$ is small compared to U.

Equation 12.17 then becomes

$$m\ddot{y} + k(x_0 + y) = -F(\alpha) \tag{12.20}$$

where we can approximate $F(\alpha)$ with the first two terms out of the Taylor's Series expansion. That is, we can write, for small α,

$$F(\alpha) \approx F(0) + \left(\frac{dF}{d\alpha}\right)\alpha \tag{12.21}$$

We use α from Equation 12.19 to get

$$F(\alpha) \approx F(0) + \frac{1}{U}\left(\frac{dF}{d\alpha}\right)\dot{y} \tag{12.22}$$

and substitute into Equation 12.20, yielding

$$m\ddot{y} + kx_0 + ky = -F(0) - \frac{1}{U}\left(\frac{dF}{d\alpha}\right)\dot{y} \tag{12.23}$$

The equilibrium condition of Equation 12.18 lets us cancel out kx_0 on the left side with $-F(0)$ on the right-hand side leaving us with, after rearrangement,

$$m\ddot{y} + \frac{1}{U}\left(\frac{dF}{d\alpha}\right)\dot{y} + ky = 0 \tag{12.24}$$

Equation 12.24 indicates that the air flow has introduced a viscous damping term (i.e. a term proportional to velocity) in the equation of motion. That would be a good thing if we could guarantee that the damping coefficient is positive. If the damping is negative, the damping force will be in the same direction of as the velocity rather than opposing it and we will get oscillations that grow without bound. The system is unstable with negative damping.

The sign on the damping will depend on whether or not $(dF/d\alpha)$ is positive or negative. There are two possibilities

$$\text{if } \left(\frac{dF}{d\alpha}\right) > 0 \text{ the system is stable}$$

$$\text{or, if } \left(\frac{dF}{d\alpha}\right) < 0 \text{ the system is unstable.}$$

We defined $F(\alpha) = L\cos\alpha + D\sin\alpha$ just after Equation 12.16. Differentiating with respect to α gives

$$\left(\frac{dF}{d\alpha}\right) = \frac{dL}{d\alpha}\cos\alpha - L\sin\alpha + \frac{dD}{d\alpha}\sin\alpha + D\cos\alpha \tag{12.25}$$

or

$$\left(\frac{dF}{d\alpha}\right) = \left[\frac{dL}{d\alpha} + D\right]\cos\alpha + \left[\frac{dD}{d\alpha} - L\right]\sin\alpha \tag{12.26}$$

For small α, $\cos\alpha \approx 1 >> \sin\alpha$, so that

$$\left(\frac{dF}{d\alpha}\right) \approx \left[\frac{dL}{d\alpha} + D\right] \tag{12.27}$$

and the system will be unstable if

$$\frac{dL}{d\alpha} + D < 0 \tag{12.28}$$

The condition for aerodynamic instability in Equation 12.28 has become known as *Den Hartog's criterion*.

If you now consider Figure 12.4, you will see that the the slope of the lift force curve with respect to the angle of attack has a large range where it is negative and the drag force is relatively small (from 45 degrees to 135 degrees) and the system will exhibit unstable oscillations in this range. This is what lies behind the *galloping* of electrical transmission lines that are covered with ice during times of freezing rain. The lines are supported elastically by the tensions in them, have mass, and their cylindrical shape becomes elongated downward by the ice hanging from them[3]. This moves them into the unstable range and steady winds can (and do) cause large amplitude vibrations that can destroy the cables and the towers that support them.

Figure 12.7 shows another effect that flow has on structures – von Karman vortex shedding. This is the case where flow over an object is sufficiently fast that vortices are shed in the wake. The vortices are alternately shed from one side of the structure or the other in a cyclic fashion. The frequency of vortex shedding can be found from

$$f_s = S\frac{V}{D} \tag{12.29}$$

where f_s is the frequency in cycles per second, V is the wind speed, D is the width of the frontal area, and S is the *Strouhal number*, a dimensionless number whose value is typically in the range of 0.15 to 0.25 for most bodies. $S = 0.2$ is typical for circular cylinders.

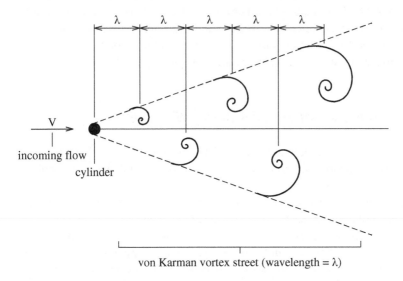

von Karman vortex street (wavelength $= \lambda$)

Figure 12.7 Representation of a von Karman vortex street in a wake.

---

3 see Sections 8.1 and 8.2 where we derived the equation of motion and found the natural frequencies for a taut string

Figure 12.8 Helical strakes on tall chimneys.

Vortex shedding is the most common cause of flow-induced vibrations. It is what causes flags to flap in the wind, for instance, as vortices are shed from the flag pole. It has also been responsible for failures of tall chimneys where the wind speed caused the frequency of vortex shedding to match a natural frequency in the structure. You will often see helical strakes on chimneys as, for example, in Figure 12.8. They are designed to disturb the regular vortex shedding so that resonance conditions can be avoided. The phenomenon is often termed *Vortex-induced Vibration (VIV)* and can cause problems in many flow/structure interactions such as those occurring in underwater pipelines and heat exchanger tubes.

12.3 Self-Excited Oscillations of Railway Wheelsets

Rail vehicles are supported at each end by *trucks* (also known as *bogies*) that are themselves supported by two *wheelsets* that roll along the rails. Figure 12.9 shows a truck and its wheelsets.

Figure 12.10 is a schematic that shows some of the parameters used when describing a wheelset. Firstly, note that the wheelset is a single rigid body comprising two tapered and flanged wheels with a solid axle connecting them. The parameters shown are the half-gauge length, b, the nominal wheel radius, r_0, and the conicity, λ. In railway parlance, the gauge is the distance between the rails, so b is half the distance between the normal points of contact of the wheels and the rails. The wheel radius varies over the width of the wheel. r_0 is the radius at the normal points of contact. The conicity is a measure of the taper of the wheels. The wheels are, in fact, a pair of back-to-back cones that have been truncated and have had material removed (see Figure 12.11).

Figure 12.9 A railway truck supported by its two wheelsets.

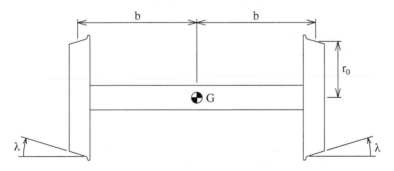

Figure 12.10 Parameters for a railway wheelset.

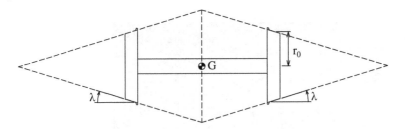

Figure 12.11 Back-to-back cones forming a railway wheelset.

The idea behind the coned wheels is that the wheelset will prefer to run in a lateral position where the radius of the wheels is the same on each side. For example, if the wheelset is somehow disturbed so that it is running off the centerline of the tracks, the wheel on the side to which it has moved will have a bigger radius than the wheel on the other side. Since the wheels are rigidly connected, they are forced to make the same number of revolutions over any distance traveled. As a result, the wheel with the larger radius will travel farther and the wheelset will be steered back toward the centerline. This self-centering effect has been employed since the earliest days of rail travel and it remains today.

We want to consider the dynamics of a wheelset as it rolls down the track at a constant speed, V. The forces acting on the wheelset arise strictly from its interaction with the rails. These forces are called *wheel-rail forces* and are related to the friction acting between the wheel and the rail. We need to quantify these forces and we start by defining the wheelset degrees of freedom and then using a kinematic analysis to see how the velocity of the contact points between the wheels and the rails varies with the degrees of freedom. Clearly, the rails have zero velocity at the contact points, so any velocity the wheels have there will lead to slipping and will result in forces being generated.

Figure 12.12 shows a top view of the wheelset in a displaced position. The dashed lines are aligned with and perpendicular to the rails. The constant forward speed, V, is shown aligned with the direction of travel and the wheelset is assumed to be traveling on straight track[4]. The degrees of freedom are: $y = $ the lateral displacement of the center of mass from the track centerline measured positively to the right and $\psi = $ the yaw angle of the wheelset measured positively in a clockwise direction.

Two coordinate systems are shown in the figure.

The $(\vec{i}_0, \vec{j}_0, \vec{k}_0)$ system is fixed in the ground with $\vec{i}_0$ aligned with the direction of travel and $\vec{j}_0$ perpendicular to the direction of travel and positive to the right. To have a right-handed coordinate system, it is therefore necessary to define $\vec{k}_0$ to be positive downwards. This coordinate system has no angular velocity.

The $(\vec{i}_1, \vec{j}_1, \vec{k}_1)$ system is fixed in the wheelset and has the angular velocity of the wheelset. That is

$$\vec{\omega}_1 = -\left(\frac{V}{r_0}\right)\vec{j}_1 + \dot{\psi}\vec{k}_1 \tag{12.30}$$

where the $\vec{k}_1$ term is clear. The $\vec{j}_1$ term requires a little explanation. It is the rolling angular velocity of the wheelset and can be derived by letting the wheelset roll without slipping

---

4 To railway people, straight track is *tangent track*

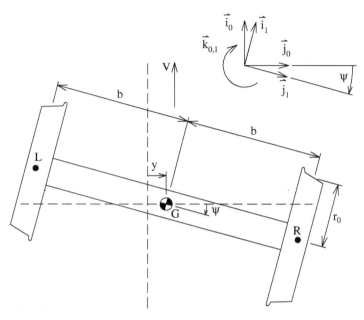

Figure 12.12 Wheelset degrees of freedom.

down the track with $\psi = 0$. The axle has velocity, V, in the forward direction and the contact points have zero velocity so that the angular velocity about a line through the axle must be V/r_0. By the right hand rule, the direction of rotation is about the negative $\vec{j}_1$ axis, thereby giving the expression shown.

We define the point, R, in Figure 12.12 to be the point of contact between the wheel and the rail on the right wheel. Its velocity can be found from

$$\vec{v}_R = \vec{v}_G + \vec{v}_{R/G} \tag{12.31}$$

where

$$\vec{v}_G = V\vec{\imath}_0 + \dot{y}\vec{\jmath}_0 \tag{12.32}$$

and

$$\vec{v}_{R/G} = \frac{d}{dt}\vec{p}_{R/G} = \dot{\vec{p}}_{R/G} + \vec{\omega}_1 \times \vec{p}_{R/G} \tag{12.33}$$

The position of R with respect to G, $\vec{p}_{R/G}$, can be written in the wheelset fixed coordinate system as

$$\vec{p}_{R/G} = b\vec{\jmath}_1 + r_R\vec{k}_1 \tag{12.34}$$

where r_R is the radius of the right wheel. r_R is only equal to r_0 if the lateral displacement, y, is zero. Otherwise we find r_R by taking the conicity, λ, into account, as follows

$$r_R = r_0 + \lambda y \tag{12.35}$$

This expression takes account of the increase in rolling radius as the wheel moves outward from the center of the track.

We can then differentiate the expression in Equation 12.34 to get

$$\vec{v}_{R/G} = \left[-\left(\frac{V}{r_0}\right)\vec{j}_1 + \dot{\psi}\vec{k}_1 \right] \times [b\vec{j}_1 + (r_0 + \lambda y)\vec{k}_1] \tag{12.36}$$

where we have taken into account the fact that the rate of change of magnitude term, $\dot{\vec{p}}_{R/G}$, is zero because all of the components of $\vec{p}_{R/G}$ are of fixed length. The result is

$$\vec{v}_{R/G} = \left[-\left(\frac{V}{r_0}\right)(r_0 + \lambda y) - \dot{\psi}b \right]\vec{i}_1 \tag{12.37}$$

We now want to transform $\vec{v}_G$ into the same coordinates as $\vec{v}_{R/G}$ so we can add them together. The relationship between the coordinates shown in Figure 12.12 is

$$\vec{i}_0 = \cos\psi\vec{i}_1 - \sin\psi\vec{j}_1 \tag{12.38}$$

and

$$\vec{j}_0 = \sin\psi\vec{i}_1 + \cos\psi\vec{j}_1 \tag{12.39}$$

Transforming $\vec{v}_G$ from Equation 12.32

$$\vec{v}_G = (V\cos\psi + \dot{y}\sin\psi)\vec{i}_1 + (-V\sin\psi + \dot{y}\cos\psi)\vec{j}_1 \tag{12.40}$$

Substituting Equations 12.37 and 12.40 into Equation 12.31 gives

$$\vec{v}_R = \left[V\cos\psi + \dot{y}\sin\psi - \left(\frac{V}{r_0}\right)(r_0 + \lambda y) - \dot{\psi}b \right]\vec{i}_1 + (-V\sin\psi + \dot{y}\cos\psi)\vec{j}_1 \tag{12.41}$$

We simplify Equation 12.41 somewhat to get

$$\vec{v}_R = \left[V\cos\psi + \dot{y}\sin\psi - V - \frac{V\lambda y}{r_0} - \dot{\psi}b \right]\vec{i}_1 + (-V\sin\psi + \dot{y}\cos\psi)\vec{j}_1 \tag{12.42}$$

Finally, for the purposes of vibrational studies, we linearize Equation 12.42 by letting y and ψ be small variations away from equilibrium. We can write $\cos\psi \approx 1$ and $\sin\psi \approx \psi$ to get

$$\vec{v}_R = \left[V + \dot{y}\psi - V - \frac{V\lambda y}{r_0} - \dot{\psi}b \right]\vec{i}_1 + (-V\psi + \dot{y})\vec{j}_1 \tag{12.43}$$

Clearly the $+V$ and the $-V$ terms in the $\vec{i}_1$ component cancel each other and the $\dot{y}\psi$ term is the product of two small variables and is therefore nonlinear and negligible. The terms in the $\vec{j}_1$ component are all linear since the forward speed, V, is a large constant that is one of the parameters of the model. We are left with

$$\vec{v}_R = \left(-\frac{V\lambda y}{r_0} - \dot{\psi}b \right)\vec{i}_1 + (-V\psi + \dot{y})\vec{j}_1 \tag{12.44}$$

If we were to repeat this analysis for the left wheel where the contact point is labeled L in Figure 12.12, there would be two differences. First, the position of L with respect to G would have $-b$ rather than the $+b$ that it has in $\vec{p}_{R/G}$ (see Equation 12.34). Secondly, the left wheel radius decreases as the wheelset moves to the right so that the equivalent expression to that in Equation 12.35 is

$$r_L = r_0 - \lambda y \tag{12.45}$$

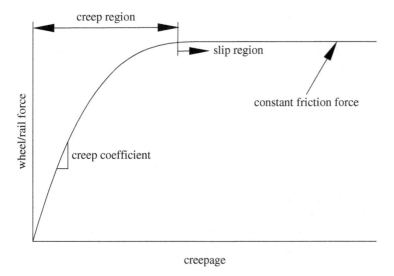

Figure 12.13 Creep forces.

As a result, we can get an expression for $\vec{v}_L$ by simply putting negative signs on b and λ in Equation 12.44, giving

$$\vec{v}_L = \left(\frac{V \lambda y}{r_0} + \dot{\psi} b \right) \vec{i}_1 + (-V\psi + \dot{y}) \vec{j}_1 \tag{12.46}$$

Equations 12.44 and 12.46 show that the points of contact between the wheel and the rail will have non-zero velocities so long as the degrees of freedom, y and ψ, and their velocities, $\dot{y}$ and $\dot{\psi}$, are not zero. It remains to quantify the forces that act on the wheels when they move over the rails.

The forces that act between the wheel and the rail cannot truly be said to be friction forces because there is no gross sliding of the wheels under normal operation. Nevertheless, the forces act in a very similar fashion to friction and, in the limiting case where sliding actually occurs, they are frictional forces as we usually understand them.

Figure 12.13 shows how the wheel rail forces vary with a variable called *creepage*. Creepage is a non-dimensional measure of the contact point velocity that we get by dividing the velocities by the forward velocity of the wheelset. The creepage of the right wheel, ξ_R, is developed by dividing Equation 12.44 by V, yielding

$$\vec{\xi}_R = \left(-\frac{\lambda y}{r_0} - \frac{\dot{\psi} b}{V} \right) \vec{i}_1 + \left(-\psi + \frac{\dot{y}}{V} \right) \vec{j}_1 \tag{12.47}$$

and, on the left, we get

$$\vec{\xi}_L = \left(\frac{\lambda y}{r_0} + \frac{\dot{\psi} b}{V} \right) \vec{i}_1 + \left(-\psi + \frac{\dot{y}}{V} \right) \vec{j}_1 \tag{12.48}$$

Creepage was first discussed by F.W. Carter[5] who was doing experiments with locomotives under load. He measured the number of revolutions that the driving wheels made as they

5 Carter, F.W., 1926. *On the action of a locomotive driving wheel*, Proceedings of the Royal Society of London, A112, pp. 151–157.

traversed a fixed length of track and discovered that the number varied with the traction that was required to pull the loads. The heavier the load, the higher the traction, and the higher the number of revolutions. If a wheel rolls without slipping, it will always take the same number of revolutions to travel the same distance. It was clear that the locomotive wheels were not "slipping" but the evidence said that neither were they "rolling without slipping". Carter coined the term "creepage" to explain the partial slip phenomenon that he observed. Today there exist highly accurate theoretical methods for explaining the phenomenon and even quantifying the magnitude of the creep forces. They are based on the concept of "contact patch" as opposed to "contact point" and include Hertzian contact stresses in elliptical contact patches. A portion of the contact patch slips and a portion doesn't during creepage. When the entire contact patch slips, the forces are friction forces as we usually understand them and the force remains constant at the coefficient of friction multiplied by the normal force once outside the creep region.

It is sufficient for the analysis that we are undertaking here to say that, in the linear region near zero creepage, the wheel/rail force is proportional to creepage and that the constant of proportionality is the *creep coefficient* shown as the slope of the curve near the origin in Figure 12.13. We define f_x to be the longitudinal creep coefficient ($\vec{\imath}_1$ direction) and f_y to be the lateral creep coefficient ($\vec{\jmath}_1$ direction).

We can then write expressions for the wheel/rail forces as

$$\vec{F}_R = -f_x \left(-\frac{\lambda y}{r_0} - \frac{\dot{\psi} b}{V} \right) \vec{\imath}_1 - f_y \left(-\psi + \frac{\dot{y}}{V} \right) \vec{\jmath}_1 \tag{12.49}$$

and

$$\vec{F}_L = -f_x \left(\frac{\lambda y}{r_0} + \frac{\dot{\psi} b}{V} \right) \vec{\imath}_1 - f_y \left(-\psi + \frac{\dot{y}}{V} \right) \vec{\jmath}_1 \tag{12.50}$$

where $\vec{F}_R$ and $\vec{F}_L$ are the forces that the rails impart to the wheels thus introducing the negative signs on the creep coefficients because these forces have to resist the creep.

We can define F_x and F_y to be

$$F_x = f_x \left(\frac{\lambda y}{r_0} + \frac{\dot{\psi} b}{V} \right) \tag{12.51}$$

and

$$F_y = f_y \left(-\psi + \frac{\dot{y}}{V} \right) \tag{12.52}$$

and put these wheel/rail forces on the free body diagram in Figure 12.14 with the directions as shown.

A lateral force balance will result in

$$m\ddot{y} = -2F_y \tag{12.53}$$

where m is the mass and $\ddot{y}$ is the lateral acceleration of the wheelset.

A moment balance about the center of mass, G, yields

$$I\ddot{\psi} = -2F_x b \tag{12.54}$$

where I is the moment of inertia about G and $\ddot{\psi}$ is the angular acceleration.

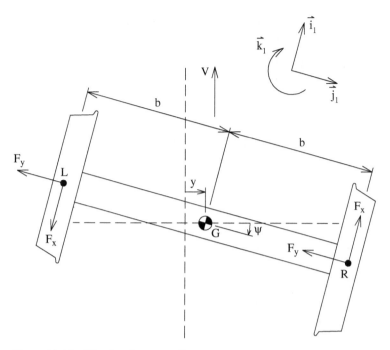

Figure 12.14 Wheelset free body diagram.

We can substitute Equations 12.51 and 12.52 into Equations 12.53 and 12.54 to get a single set of matrix equations

$$
\begin{bmatrix} m & 0 \\ 0 & I \end{bmatrix} \begin{Bmatrix} \ddot{y} \\ \ddot{\psi} \end{Bmatrix} + \begin{bmatrix} \frac{2f_y}{V} & 0 \\ 0 & \frac{2f_x b^2}{V} \end{bmatrix} \begin{Bmatrix} \dot{y} \\ \dot{\psi} \end{Bmatrix} + \begin{bmatrix} 0 & -2f_y \\ \frac{2f_x b\lambda}{r_0} & 0 \end{bmatrix} \begin{Bmatrix} y \\ \psi \end{Bmatrix} = \begin{Bmatrix} 0 \\ 0 \end{Bmatrix}
$$

(12.55)

There are several things that can be done with Equation 12.55. The first is to assume that the forward speed is relatively low and that motions are gentle enough to make the inertial terms negligible with respect to the wheel/rail forces. We can remove the acceleration terms from the equations of motion, giving the following set of first-order differential equations

$$
\begin{bmatrix} \frac{1}{V} & 0 \\ 0 & \frac{b}{V} \end{bmatrix} \begin{Bmatrix} \dot{y} \\ \dot{\psi} \end{Bmatrix} + \begin{bmatrix} 0 & -1 \\ \frac{\lambda}{r_0} & 0 \end{bmatrix} \begin{Bmatrix} y \\ \psi \end{Bmatrix} = \begin{Bmatrix} 0 \\ 0 \end{Bmatrix}
$$

(12.56)

where the first equation has been divided throughout by $2f_y$ and the second by $2f_x b$.

We make the usual assumption that solutions to Equation 12.56 are exponential so that we can write

$$
\begin{Bmatrix} y \\ \psi \end{Bmatrix} = \begin{Bmatrix} Y \\ \Psi \end{Bmatrix} e^{\mu t}
$$

(12.57)

where Y and Ψ are unknown amplitudes and μ is an unknown parameter. Differentiating Equation 12.57 and substituting the result into Equation 12.56 gives

$$\begin{bmatrix} \frac{\mu}{V} & -1 \\ \frac{\lambda}{r_0} & \frac{b\mu}{V} \end{bmatrix} \begin{Bmatrix} Y \\ \Psi \end{Bmatrix} e^{\mu t} = \begin{Bmatrix} 0 \\ 0 \end{Bmatrix} \tag{12.58}$$

Equation 12.58 has a zero right-hand side and an unknown in the coefficient matrix so it represents an eigenvalue problem and we can extract the characteristic equation by setting the determinant of the coefficient matrix to zero, resulting in

$$\left(\frac{\mu}{V} \right) \left(\frac{b\mu}{V} \right) - \left(\frac{\lambda}{r_0} \right) (-1) = 0 \tag{12.59}$$

or

$$\left(\frac{b}{V^2} \right) \mu^2 + \left(\frac{\lambda}{r_0} \right) = 0 \tag{12.60}$$

from which

$$\mu^2 = - \left(\frac{\lambda}{r_0} \right) \left(\frac{V^2}{b} \right) = -V^2 \left(\frac{\lambda}{br_0} \right) \tag{12.61}$$

Equation 12.61 lets us solve for μ, yielding

$$\mu = \pm iV \sqrt{\frac{\lambda}{br_0}} \tag{12.62}$$

where i is the square root of minus one. μ is an imaginary number indicating that the assumed exponential solution is actually a harmonic motion (from Euler's Identity) and that the wheelset response to an initial disturbance will be oscillatory with frequency, ω, given by

$$\omega = V \sqrt{\frac{\lambda}{br_0}} \tag{12.63}$$

While the frequency of motion varies with speed as is to be expected. However, we get the interesting result that the wavelength of the motion does not. If we consider the period of the motion to be T, we have, by definition,

$$\omega T = 2\pi \tag{12.64}$$

The wavelength of the motion along the track, L_w, is defined to be the distance traveled during one period of motion, or

$$L_w = VT \tag{12.65}$$

We can write $T = 2\pi/\omega$ from Equation 12.64 and substitute it into Equation 12.65, giving

$$L_w = \frac{2\pi V}{\omega} \tag{12.66}$$

Then, we can substitute ω from Equation 12.63 to arrive at

$$L_w = 2\pi V \left(\frac{1}{V} \sqrt{\frac{br_0}{\lambda}} \right) \tag{12.67}$$

or

$$L_w = 2\pi \sqrt{\frac{br_0}{\lambda}}$$

(12.68)

L_w is called the *kinematic wavelength*. Under the assumed conditions of gentle motion where inertial effects can be ignored, the wheelset oscillates down the track with a wavelength independent of speed.

We now consider the case where inertial effects are retained and we use the full equations of motion from Equation 12.55. We again assume exponential solutions as in Equation 12.57 and the result is

$$
\begin{bmatrix} m\mu^2 + \frac{2f_y}{V}\mu & -2f_y \\[2mm] \frac{2f_x b\lambda}{r_0} & I\mu^2 + \frac{2f_x b^2}{V}\mu \end{bmatrix}
\begin{Bmatrix} Y \\ \Psi \end{Bmatrix} e^{\mu t} =
\begin{Bmatrix} 0 \\ 0 \end{Bmatrix}
$$

(12.69)

The characteristic equation formed from Equation 12.69 is

$$\mu^4 [mI] + \mu^3 \left[\frac{2mf_x b^2}{V} + \frac{2If_y}{V} \right] + \mu^2 \left[\frac{4f_x f_y b^2}{V^2} \right] + \mu [0] + \frac{4f_x f_y b\lambda}{r_0} = 0$$

(12.70)

It is virtually impossible to extract the roots of this fourth-order characteristic equation analytically. We therefore rely on another method for looking at stability of the equilibrium state that is constant speed motion along the center line of straight track – the method of Routh.

The Routh stability criterion is a method for determining system stability from the n-th order characteristic equation of the form:

$$a_n \lambda^n + a_{n-1} \lambda^{n-1} + \cdots + a_1 \lambda + a_0 = 0$$

(12.71)

From the coefficients of the characteristic equation, we form a *Routh Table* as shown in Table 12.1. The first two rows of the table contain the coefficients of the characteristic equation and are padded with zeroes at the end. We find the coefficients b_i and c_i from 2×2 determinants involving the coefficients in the two rows just preceding them. That is,

$$b_1 = \frac{a_{n-1} a_{n-2} - a_n a_{n-3}}{a_{n-1}} \quad b_2 = \frac{a_{n-1} a_{n-4} - a_n a_{n-5}}{a_{n-1}} \quad \cdots$$

(12.72)

$$c_1 = \frac{b_1 a_{n-3} - a_{n-1} b_2}{b_1} \quad c_2 = \frac{b_1 a_{n-5} - a_{n-1} b_3}{b_1} \quad \cdots$$

Table 12.1 The Routh Table.

λ^n	a_n	a_{n-2}	a_{n-4}	$\cdots$	0
λ^{n-1}	a_{n-1}	a_{n-3}	a_{n-5}	$\cdots$	0
λ^{n-2}	b_1	b_2	b_3	$\cdots$	
λ^{n-3}	c_1	c_2	c_3	$\cdots$	
$\vdots$	$\vdots$	$\vdots$	$\vdots$	$\vdots$	

Each new element generated for the table depends on the leading elements of the preceding two rows and on the two elements just to the right on the same two rows. For instance, the calculation of c_2, the element in row 4 and column 2, uses a_{n-1} and b_1, the two leading elements in rows 2 and 3, and a_{n-5} and b_3, the two elements in column 3 of rows 2 and 3. The pattern is easy to follow and the table is continued horizontally and vertically until only zeroes are obtained.

A statement on the stability of the equilibrium state being analyzed for the system can be made from consideration of the first column of the Routh table as follows – *all of the roots of the characteristic equation have negative real parts if and only if the elements of the first column of the Routh table have the same sign. Otherwise, the number of roots with positive real parts is equal to the number of sign changes in the first column.*

Consider the characteristic equation to be

$$a_4\mu^4 + a_3\mu^3 + a_2\mu^2 + 0\mu + a_0 = 0 \tag{12.73}$$

where the coefficients, a_i, represent the coefficients in Equation 12.70 and it is noted that the coefficient of the linear term is zero. The Routh table for this characteristic equation is shown in Table 12.2.

Looking at the coefficients in Equation 12.70, you can see that they are all positive combinations of the system parameters except in the case where the coefficient is zero. Now looking at the first column of the Routh table you can see that there are two sign changes. The first is the transition from the λ^2 row to the λ^1 row, where the change is from a_2 to $-a_3a_0/a_2$. The second sign change is from the λ^1 row to the λ^0 row, where the change is from $-a_3a_0/a_2$ to a_0.

The result is that there are two roots with positive real parts so the wheelset is unstable. The kinematic oscillations that we discussed earlier actually grow in amplitude when we include mass in the model. The conclusion is that mass has a destabilizing influence on

Table 12.2 The Routh Table for the wheelset.

λ^4	a_4	a_2	a_0	0
λ^3	a_3	0	0	
λ^2	a_2	a_0		
λ^1	$-\frac{a_3a_0}{a_2}$	0		
λ^0	a_0			

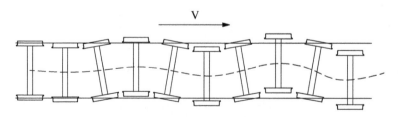

Figure 12.15 Wheelset instability.

wheelset stability. Figure 12.15 shows the unstable motion which is called *wheelset hunting* in the industry.

There are many more aspects to the dynamic performance of rail vehicles than this simple stability problem but "hunting" persists to this day in all rail vehicles. It can be shown that adding a suspension (i.e. stiffnesses that connect the wheelset to another body that moves down the track) can effectively stabilize the wheelset but the "body" that it attached to must be supported at the other end by another wheelset (see the truck in Figure 12.9) and then we find that the wheelsets and truck together can also experience hunting. And so it goes as you construct a carbody for the rail vehicle that is supported by a truck at either end; you find that the entire system still hunts above a certain critical speed. Designing rail vehicles for stability is still very much in the forefront of activity in the rail industry.

12.4 What is a Rigid Body Mode?

There are modes of vibrating systems where masses demonstrate a tendency to move away from their equilibrium positions without restraint. In these cases, although there are elements providing restoring forces to the system, the elements do not resist the motion associated with the rigid body mode. The defining characteristic of a rigid body mode is that it has a natural frequency of zero, indicating that there are no oscillations – once it moves away, it never comes back.

Consider the system shown in Figure 12.16. The system has five components, as follows,

- a thin, uniform, rigid rod with mass, m, and length, 2ℓ.

Figure 12.16   System with a rigid body mode.

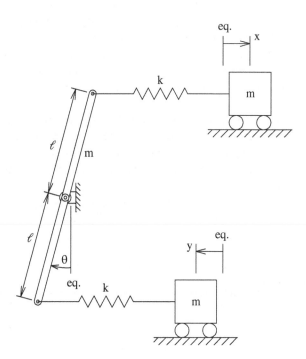

- two rigid translating masses, m and m.
- two linear springs, k and k.

Using the three degrees of freedom shown, x, y, and θ, the equations of motion can be shown to be

$$\begin{bmatrix} m & 0 & 0 \\ 0 & m & 0 \\ 0 & 0 & m\ell^2/3 \end{bmatrix} \begin{Bmatrix} \ddot{x} \\ \ddot{y} \\ \ddot{\theta} \end{Bmatrix} + \begin{bmatrix} k & 0 & -k\ell \\ 0 & k & -k\ell \\ -k\ell & -k\ell & 2k\ell^2 \end{bmatrix} \begin{Bmatrix} x \\ y \\ \theta \end{Bmatrix} = \begin{Bmatrix} 0 \\ 0 \\ 0 \end{Bmatrix} \tag{12.74}$$

With a little effort, it can be shown that the three natural frequencies and their corresponding mode shapes are

$$\omega_1 = 0 \text{ with } \begin{Bmatrix} X \\ Y \\ \Theta \end{Bmatrix}_1 = \begin{Bmatrix} 1 \\ 1 \\ 1/\ell \end{Bmatrix} \tag{12.75}$$

$$\omega_2 = \sqrt{k/m} \text{ with } \begin{Bmatrix} X \\ Y \\ \Theta \end{Bmatrix}_2 = \begin{Bmatrix} 1 \\ -1 \\ 0 \end{Bmatrix} \tag{12.76}$$

$$\omega_3 = \sqrt{7k/m} \text{ with } \begin{Bmatrix} X \\ Y \\ \Theta \end{Bmatrix}_3 = \begin{Bmatrix} 1 \\ 1 \\ -6/\ell \end{Bmatrix} \tag{12.77}$$

Figure 12.17 shows the three modes, one in each column, with time, or, more precisely, fractions of a cycle shown on the vertical. This way of presenting mode shapes lets the reader scan from top to bottom while imagining an animation of the motion of the system.

The second and third modes are typical of vibrating systems. The second mode sees no rotation of the rod so the carts oscillate back and forth with equal amplitudes[6] and a natural frequency exactly equal to what it would be if they were independent 1DOF systems. The third mode has the carts traveling in opposite directions with relatively large rotations of the rod.

The first mode is the *rigid body mode* that is the topic of this section. This is a mode where the top mass moves, the bar rotates, and the bottom mass moves in a manner that keeps all spring displacements equal to zero so there can be no restoring forces. The figure shows the motion for a clockwise rotation of the rod but it would work just as well for a counterclockwise rotation.

If you are modeling systems and your analysis finds a rigid body mode (i.e. there is a zero natural frequency) when you don't expect one, then you have probably forgotten to look after an elastic element that ties the system to ground.

---

6 At first glance it may appear that the two masses moving in the same direction at the same time is at odds with the mode shape that has $X = 1$ while $Y = -1$. If you refer back to Figure 12.16, you will see that X and Y are defined to be positive in opposite directions and the second mode shape has their ratio as -1, so they are physically moving in the same direction at the same time.

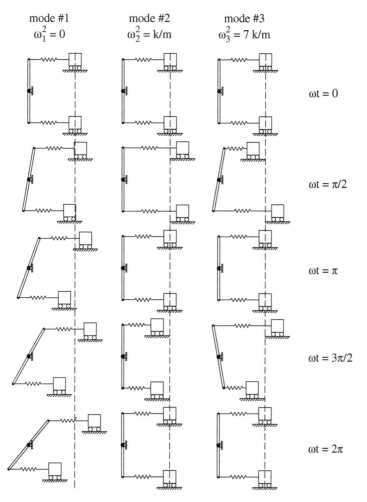

mode #1
$\omega_1^2 = 0$

mode #2
$\omega_2^2 = k/m$

mode #3
$\omega_3^2 = 7\ k/m$

$\omega t = 0$

$\omega t = \pi/2$

$\omega t = \pi$

$\omega t = 3\pi/2$

$\omega t = 2\pi$

Figure 12.17 The three modes.

12.5 Why Static Deflection is Very Useful

Throughout the material that we have covered, we have been considering systems that were made to look like a mass sitting on a spring. There were often other things like damping and complex geometry but we always came back to the model shown in Figure 12.18 where we said the mass was the effective mass of the system and the stiffness was the effective stiffness of the system. Early on, we determined that a system like this, when disturbed from a stable equilibrium state, would oscillate at only one frequency. We called the the natural frequency, ω_n, where

$$\omega_n = \sqrt{\frac{k}{m}} \tag{12.78}$$

Figure 12.18 A spring/mass system.

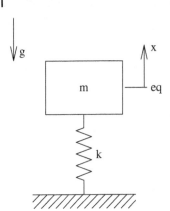

If you step back a bit and write the static force balance equation that led to the determination of the equilibrium state, you find

$$mg = k\Delta \tag{12.79}$$

where Δ is the static deflection of the spring, a measure of how much the spring changed in length when the mass was placed on it.

We can manipulate Equation 12.79 to get

$$\frac{k}{m} = \frac{g}{\Delta} \tag{12.80}$$

and, substituting this into Equation 12.78, we find

$$\omega_n = \sqrt{\frac{g}{\Delta}} \tag{12.81}$$

This seems unremarkable until you realize that we can find the natural frequencies of systems by knowing nothing about them except how far they settled when they were placed on the structure that is supporting them.

For example, if you look at a large air compressor in a shop, it will be mounted to the floor on springs. If you look around for spare parts near the compressor, you are likely to find one of the suspension springs. Measure the length of that spring and you have the undeflected length. Then, measure the length of a spring under the compressor. The difference in these lengths is Δ. With this you can calculate the natural frequency of the vertical vibration mode without knowing the mass of the compressor or the stiffness of the spring. It doesn't matter how many springs are supporting the compressor, both the static deflection and the natural frequency change as the number of springs change but Equation 12.81 will still be correct.

If you walk around a parking lot, you will see cars sitting still. They are in equilibrium with gravity pulling them down and suspension and tire preloads holding them up. If you push down on the fender of a car and then let it go, you will have displaced it away from equilibrium and it will oscillate briefly and then stop due to the action of the shock absorbers and to the dry friction that is inherent in any mechanism. You may wonder how stiff suspension springs on cars are. You can get a good estimate by looking at frequencies that people find to be comfortable.[7] That frequency, for vertical vibrations of vehicles, is around 1.5 Hz. Using

---

7 More to the point, we look at frequencies that people don't like and then choose to design to a frequency that is the least uncomfortable.

Equation 12.81, you can work out fairly quickly that the static deflection of the suspension should be about 11 cm. This result is independent of the details of the design. You need to come up with something that deflects 11 cm when the chassis of the car and the payload, usually passengers, are placed upon it. Figure 2.2 in Section 2.1 shows a fairly typical "double wishbone" suspension. There is discussion there about how this structure has a single degree of freedom that we can take to be the vertical displacement of the tire so long as the chassis is considered to be ground. Basically, this corner of the vehicle is tasked with supporting one quarter, more or less, of the weight of the vehicle. To get an approximation to the stiffness of the inclined spring shown there, we need to look at this as a structure that will have a vertical load applied to the tire and that the tire will move 11 cm for an applied load equal to one quarter of the weight of the vehicle.

Figure 12.19 shows the model we would use to find the effective stiffness of the suspension. You will see that many extraneous things have been removed if you compare this to Figure 2.2. It's not that the removed elements are unimportant to the design but only that, for the small motions we are considering, they have no effect. Essentially, we are looking at a lever that deflects an inclined spring when the end of the lever (point C) moves up or down through a distance x. It is fairly simple, and left as an exercise, to show that the stiffness term in the equation of motion will be

$$k \left[\left(\frac{a}{a+b} \right)^2 \cos^2\theta \right] x \qquad (12.82)$$

giving an effective stiffness of

$$k_{eff} = k \left[\left(\frac{a}{a+b} \right)^2 \cos^2\theta \right]$$

for vertical displacements, x, of the center of the tire.

Now, all you have to do is say that a vertical force, F, equal to one quarter of the weight of the vehicle is applied to the point C, and then let $F = k_{eff}x_o$, where x_o is the static deflection. Let x_o be 11 cm and solve for k. This will be an excellent first approximation for the stiffness

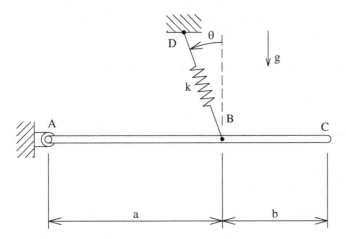

Figure 12.19 The model of a double wishbone suspension.

you need in the suspension and you are in a good position to proceed with suspension optimization.

Exercises

12.1 The figure shows two carts of mass m and $2m$, respectively. They are connected to each other by two springs, each of stiffness k. This is the same system that was analyzed in Exercise 7.3, except that the larger cart was also connected to the ground by a spring of stiffness k there. That stiffness is now absent, so there is no longer a connection to ground.

(a) Use absolute coordinates, x_1 and x_2, as the degrees-of-freedom where x_1 measures the displacement of the larger, supporting cart and x_2 measures the displacement of the small cart away from their static equilibrium positions. Derive the equations of motion and find the two natural frequencies and their corresponding mode shapes.

(b) You should have found a rigid body mode in part (a). What property of the stiffness matrix indicates whether or not there will be a rigid body mode?

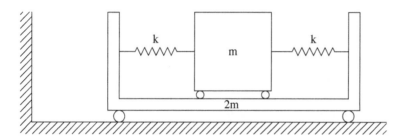

Figure E12.1

12.2 We revisit Exercise 1.4 with two methods of finding the natural frequency.
An open cylindrical container with a weight at its bottom is placed in the ocean. The cylinder sinks until buoyancy forces equal the total weight and then floats upright in equilibrium.

(a) Draw a FBD of the cylinder just after it is placed in the water. Use a DOF, y, to indicate the distance the bottom of the cylinder has traveled below the free surface of the water. There will be a buoyancy force acting upward on the cylinder. This force is due to the water pressure a distance y under the surface. That pressure is equal to $\rho g y$, where ρ is the density of water and g is the acceleration due to gravity. Assume a cross-sectional area of A for the cylinder.

(b) Use Newton's Laws to write the equation of motion for the cylinder.

(c) Find the equilibrium condition for the cylinder and solve for y_o, the equilibrium value of y.

(d) Let there be small a small motion, x, away from equilibrium so that $y = y_o + x$. Write the free vibration equation of motion in terms of x and find the undamped natural frequency.

(e) Show that the undamped natural frequency just found is equal to the square root of g divided by the static displacement, y_o.

12.3 This is Exercise 3.7, with the spring now contributing a mass as well as a stiffness. Derive expressions for the undamped natural frequency and the damping ratio for the system shown. Treat the T-shaped structure as being rigid and massless. The mass, m, is a particle. The spring has a mass of $3m$.

Figure E12.3

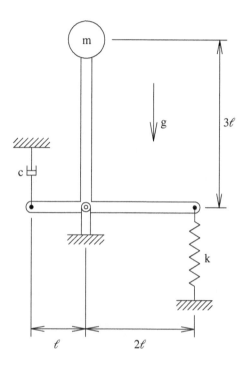

Appendix A

Least Squares Curve Fitting

This appendix is a short review of Least Squares curve fitting in aid of enhancing the understanding of Fourier Transforms in Chapter 11.

Consider the three data points, (x_n, y_n), shown in Table A.1. We want to define a function that will best approximate these data.

There is a class of curve fitting methods called *Trial Functions with Undetermined Coefficients*. To use one of these methods, we first define a set of functions, $y_i(x); i = 1, r$, that we think are good fits to the data we have and then generate an approximation by adding together these functions multiplied by a set of coefficients, a_i, which we later determine in order to get the best possible fit to the data. That is, we let the approximation be

$$y_A(x) = \sum_{i=1}^{r} a_i y_i(x) \tag{A.1}$$

and then use one of several possible methods to find the a_i. Here we restrict ourselves to using the *Method of Least Squares*.

Consider first the case where we attempt to fit a straight line to the three points. That is, we define an approximation, $y_A(x)$, as the sum of the constant function, $y_1(x) = 1$, and the linear function, $y_2(x) = x$, multiplied by undetermined coefficients, a_1 and a_2, yielding

$$y_A(x) = a_1 + a_2 x \tag{A.2}$$

The goal is to find the "best" possible values of a_1 and a_2. The "best" values are those that minimize the error at the three data points, where we define error as the difference between $y_A(x_n)$ and y_n. The error at the n-th data point is then

$$e_n = (a_1 + a_2 x_n) - y_n \tag{A.3}$$

We can't minimize the errors as shown in Equation A.3 since they have no lower limit. Very large negative errors would be deemed to be inferior to even larger negative errors in any minimization attempt.

The solution to this problem is to define a function of the errors that is always positive and then minimize that. The Least Squares Method uses the squared errors as the function to minimize. We therefore define a function, J, as

$$J = \frac{1}{2} \sum_{n=1}^{N} (e_n)^2 = \frac{1}{2} \sum_{n=1}^{N} (a_1 + a_2 x_n - y_n)^2 \tag{A.4}$$

where N is the number of data points.

Introduction to Mechanical Vibrations, First Edition. Ronald J. Anderson.
© 2020 John Wiley & Sons Ltd. Published 2020 by John Wiley & Sons Ltd.
Companion website: www.wiley.com/go/anderson/introduction-to-vibrations

Table A.1 Sample data points.

n	x_n	y_n
1	1	9
2	3	31
3	7	123

To minimize J, we simply take the partial derivatives of J with respect to the undetermined coefficients, in this case a_1 and a_2, and ensure that they are both zero. If we think of a three-dimensional plot with J plotted vertically against a_1 and a_2 axes, it would appear as a bowl with the two partial derivatives being simultaneously zero at the bottom of the bowl where J is a minimum.

The partial derivatives are

$$\frac{\partial J}{\partial a_1} = \sum_{n=1}^{3} (a_1 + a_2 x_n - y_n)(1)$$

$$= a_1 \left[\sum_{n=1}^{3} (1) \right] + a_2 \left[\sum_{n=1}^{3} (x_n) \right] - \left[\sum_{n=1}^{3} (y_n) \right] \tag{A.5}$$

and

$$\frac{\partial J}{\partial a_2} = \sum_{n=1}^{3} (a_1 + a_2 x_n - y_n)(x_n)$$

$$= a_1 \left[\sum_{n=1}^{3} (x_n) \right] + a_2 \left[\sum_{n=1}^{3} (x_n^2) \right] - \left[\sum_{n=1}^{3} (x_n y_n) \right] \tag{A.6}$$

Substituting the values for x_n and y_n from Table A.1 results in the set of linear algebraic equations

$$\begin{bmatrix} 3 & 11 \\ 11 & 59 \end{bmatrix} \begin{Bmatrix} a_1 \\ a_2 \end{Bmatrix} = \begin{Bmatrix} 163 \\ 963 \end{Bmatrix} \tag{A.7}$$

with the solution yielding

$$\begin{Bmatrix} a_1 \\ a_2 \end{Bmatrix} = \begin{Bmatrix} -17.42 \\ 19.57 \end{Bmatrix} \tag{A.8}$$

so that the Least Squares linear approximation is

$$y_A(x) = -17.42 + 19.57x \tag{A.9}$$

which is the straight line plotted on Figure A.1, where the three dots are the points from Table A.1.

Also plotted on Figure A.1 is a dashed line that goes through all three data points exactly. This is the result of using the Least Squares Method with three undetermined coefficients. That is, let the approximation be

$$y_A(x) = a_1 + a_2 x + a_3 x^2 \tag{A.10}$$

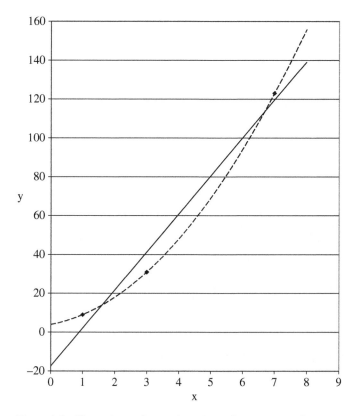

Figure A.1 Three data points and two Least Squares curve fits.

Formulating J and setting the three partial derivatives to zero give the set of equations

$$\begin{bmatrix} 3 & 11 & 59 \\ 11 & 59 & 371 \\ 59 & 371 & 2483 \end{bmatrix} \begin{Bmatrix} a_1 \\ a_2 \\ a_3 \end{Bmatrix} = \begin{Bmatrix} 163 \\ 963 \\ 6315 \end{Bmatrix} \tag{A.11}$$

Solving Equation A.11 yields

$$y_A(x) = 4 + 3x + 2x^2 \tag{A.12}$$

which is the function used to generate the data points in the beginning.

The fact that the Method of Least Squares will fit a curve that goes through all the data points if the number of undetermined coefficients is equal to the number of data points is relied upon in Chapter 11, where the trial functions are sines and cosines and the result is the Discrete Fourier Transform.

Appendix B

Moments of Inertia

In any two-dimensional problem[1] with rotational degrees of freedom, there is a need to know the moments of inertia of the bodies in the system. In systems where masses are modelled as particles, the determination of the inertia properties is simply a matter of summing particle masses multiplied by products of their distances from a reference point. In rigid bodies the number of particles is infinite and the summations become integrals, as follows.

Moments of inertia are defined about a reference point O on the axis of rotation of the body as:

Moment of Inertia about the x-axis: $\quad I_{xx_O} = \displaystyle\int_{body} (y^2 + z^2)dm$

Moment of Inertia about the y-axis: $\quad I_{yy_O} = \displaystyle\int_{body} (x^2 + z^2)dm$

Moment of Inertia about the z-axis: $\quad I_{zz_O} = \displaystyle\int_{body} (x^2 + y^2)dm$

There are several points to note about moments of inertia. First, it is clear that moments of inertia are always positive due to the fact that the integrals involve the sums of squares of distances and squared numbers are always positive. Notice that, by Pythagoras' Theorem, these sums of squares are simply the square of the distance from the particle to the reference point in the plane perpendicular to the axis about which the rotation is being considered. Also notice that adding mass to a body always increases the moment of inertia unless the mass is added at the reference point.

It is relatively simple to show that, for a rigid body, the sum of the any two moments of inertia must be greater than the third. Simply add together I_{xx} and I_{yy} to show[2]:

$$I_{xx} + I_{yy} = \int_{body} (y^2 + z^2)dm + \int_{body} (x^2 + z^2)dm$$

$$= \int_{body} (x^2 + y^2 + 2z^2)dm$$

---

1 Three-dimensional dynamic models also require *products of inertia*, but as these rarely enter into linear vibrational analysis, they are not discussed here.
2 The reference point O is dropped from the notation for brevity.

Introduction to Mechanical Vibrations, First Edition. Ronald J. Anderson.
© 2020 John Wiley & Sons Ltd. Published 2020 by John Wiley & Sons Ltd.
Companion website: www.wiley.com/go/anderson/introduction-to-vibrations

$$= \int_{body} (x^2 + y^2)dm + \int_{body} 2z^2 dm$$

$$= I_{zz} + \int_{body} 2z^2 dm > I_{zz} \tag{B.1}$$

Clearly this statement could be written for the sum of any two moments of inertia with the result that that sum must be greater than the third moment of inertia. This is a useful check on moments of inertia, especially those that are provided to the analyst by someone else. If they fail this test, then they are incorrect.

The dimensions of moments of inertia are mass times distance squared. In SI units this will be $kg \cdot m^2$.

Because the moment of inertia is the product of mass and distance squared, it has become customary and convenient to specify the moment of inertia of a body through the use of the mass of the body and a characteristic length called the **radius of gyration,** k_O, about the reference point. The radius of gyration is defined as:

$$k_O = \sqrt{\frac{I_O}{m}} \tag{B.2}$$

where m is the total mass of the rigid body and I_O is the body's moment of inertia about the reference point for the plane of motion being considered. Equation B.2 can be manipulated to read:

$$I_O = mk_O^2 \tag{B.3}$$

which can be interpreted as saying that the radius of gyration is the distance that a particle having the same mass as the rigid body would need to be from the reference point to have the same moment of inertia as the rigid body. The inertial properties of a rigid body are therefore often specified to an analyst as simply a mass and a distance. The actual moment of inertia is reconstructed using Equation B.3.

B.1 Parallel Axis Theorem for Moments of Inertia

The integration required to derive an expression for the moment of inertia is often tedious and to be avoided to the extent possible. Undergraduate textbooks on dynamics usually include extensive tables of moments of inertia for rigid bodies of various shapes. In constructing the tables, the authors have had to choose a reference point for the integration and the point chosen is most often the center of mass of the body, G. The **Parallel Axis Theorem** allows the analyst to use the moment of inertia about the center of mass to calculate the moment of inertia about a parallel axis passing through another point without performing any integrations.

Figure B.1 shows a rigid body rotating in a plane with the center of mass, G, and another reference point, O, shown. The distance from O to G is u in the x-direction any v in the y-direction or simply $d = \sqrt{u^2 + v^2}$.

I_{zz} can be calculated in this plane since all rotations are around the z-axis. The moment of inertia about the center of mass is:

$$I_{zz_G} = \int_{body} (x^2 + y^2)dm \tag{B.4}$$

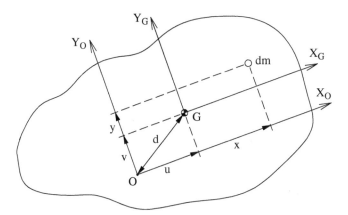

Figure B.1   Parallel Axis Theorem.

and the moment of inertia about the reference point O is:

$$I_{zz_O} = \int_{body} [(u+x)^2 + (v+y)^2]dm \tag{B.5}$$

Equation B.5 can be expanded and written as[3] :

$$I_{zz_O} = \int [(u^2 + 2ux + x^2) + (v^2 + 2vy + y^2)]dm \tag{B.6}$$

which can be rearranged to be:

$$I_{zz_O} = \int (u^2 + v^2)dm + \int (x^2 + y^2)dm + \int (2ux + 2vy)dm \tag{B.7}$$

where we can recognize that $u^2 + v^2 = d^2$, that the second integral is the moment of inertia about G (see Equation B.4) and that u and v are constants so they can be factored out of the last integral. The result is:

$$I_{zz_O} = d^2 \int dm + I_{zz_G} + 2u \int xdm + 2v \int ydm \tag{B.8}$$

The first integral term in Equation B.8 is clearly just the total mass, m, of the rigid body. The two other integrals are both definitions of the location of the center of mass with respect to the origin for the integral, G, and so define the location of the center of mass with respect to itself. Both of these integrals are therefore zero and the Parallel Axis Theorem becomes:

$$I_{zz_O} = I_{zz_G} + md^2 \tag{B.9}$$

B.2   Moments of Inertia for Commonly Encountered Bodies

Tables of moments of inertia are easy to find in undergraduate textbooks on Dynamics. Some common moments of inertia are given here for quick reference.

---

3 The integrals are assumed to be over the entire body for the remainder of this discussion so the word *body* can be omitted from the integral signs to improve readability.

Thin Uniform Rod: mass $= m$, length $= \ell$

$$I_G = \frac{1}{12} m\ell^2$$

Uniform Disk: about an axis perpendicular to the plane in which it lies, mass $= m$, radius $= r$

$$I_G = \frac{1}{2} mr^2$$

Thin Uniform Disk: about a radial axis, mass $= m$, radius $= r$

$$I_G = \frac{1}{4} mr^2$$

Uniform Rectangular Block: mass $= m$, base $= b$, height $= h$

$$I_G = \frac{1}{12} m(b^2 + h^2)$$

Index

Introduction to Mechanical Vibrations, First Edition. Ronald J. Anderson.
© 2020 John Wiley & Sons Ltd. Published 2020 by John Wiley & Sons Ltd.
Companion website: www.wiley.com/go/anderson/introduction-to-vibrations